B.F. Gray

Electronics 1: Systems and practice

Longman London and New York

Longman Group Limited London

*Associated companies, branches and representatives
throughout the world*

*Published in the United States of America
by Longman Inc., New York*

© Longman Group Limited 1977

All rights reserved. No part of this publication may be
reproduced, stored in a retrieval system, or transmitted
in any form or by any means, electronic, mechanical,
photocopying, recording, or otherwise, without the
prior permission of the Copyright owner.

First published 1977

Library of Congress Cataloging in Publication Data

Gray, Bernard Francis, 1924–
   Electronics 1: systems and practice.

   (Longman technician series)
   Includes index.
   1. Electronics. I. Title.
TK7815.G698   621.281   77-21567
ISBN 0-582-41131-9

Printed in Great Britain by
Richard Clay (The Chaucer Press) Ltd,
Bungay, Suffolk

Longman Technician Series

Electrical and Electronic Engineering

General Editor — Electrical and Electronic Engineering
I. McKenzie Smith, B.Sc.(Hons.), Dip.A.Ed., C.Eng., M.I.E.E., M.I.E.R.E., F.I.T.E.
*Head of the Department of Electrical Engineering,
Stow College of Engineering, Glasgow*

Books to be published in Electrical and Electronic Engineering sector:

Electrical drawing and communication   E. J. Pollard, I. McKenzie Smith and J. McMaster
Electrical principles and measurements Volume 1   I. McKenzie Smith
Electrical principles and measurements Volume 2   I. McKenzie Smith
Electronics 2   **B. F. Gray**
Electronics 3   **B. F. Gray**

Preface

This is a book written specifically for students embarking on the new Technician Education Council courses. It is a Level I text and therefore assumes only elementary mathematical knowledge and does not in any way make demands on the ability of students in this field.

While the book follows the guide syllabus issued by the T.E.C., opportunity is taken here and there to expand on the bare bones of the published syllabus. Two further volumes are planned in this subject area — again following the syllabus produced by the A2 Programme Committee of the T.E.C.

Students in the Electronics, Telecommunications and Communication courses should find the text very acceptable in both content and treatment. SI units are employed and preferred British Standard symbols are used. Where there are symbols in common practice which do not conform with BS specification (e.g. logic symbols) the alternatives are given.

Each chapter has an 'end test', sometimes in the form of multiple choice questions and sometimes in the form of the more usual numerical example. Occasionally there are instructions to write brief essays and invitations to read a little more than the words in the text. Teachers of the subject should find this approach useful if continuous assessment methods are employed.

Again my thanks are extended to Marion Dance, who typed the manuscript.

B.F.G.
Harpenden, 1977

We are grateful to the following for permission to reproduce copyright material: Avo Limited for Fig. 8.10; EITB for Fig. 2.4; *Electricity Times* for Fig. 2.5; Hawker Siddeley Dynamics for Fig. 4.13; Mullard Limited for Fig. 4.11; Thomas Nelson and Sons Limited for Figs. 6.1, 6.2, 6.25, 8.1, 8.2, 8.3, and 8.4 which were taken from B. F. Gray's *Electrical Engineering Principles*, 1970; R.S. Components Limited for Table 7.2; and Paul Brierley for the cover photograph.

Contents

Preface and Acknowledgements   iv

Chapter 1   The start and expansion of electronics   1
Early discoveries   2
The development of electronics,
   (i) up to the beginning of the
      Second World War   6
   (ii) during the Second World War   7
   (iii) during the post-war period   7
The development of telecommunications from
   the early nineteenth century   9
Professional institutions   11

Chapter 2   The electronics industry   17
Early days   17
The Marconi Company   18
Industrial organisation today   19
The present day electronics industry   20
The organisation of a typical electronics firm   22
Safety aspects   26

Chapter 3   The electric circuit   32
Ohm's law and its application to d.c. circuits   32
Resistors in series and parallel   35
Series-parallel arrangements   36
Resistance of different materials and the effect
   of temperature   38
Linear and non-linear resistors   40

Chapter 4   Electronic systems   46

The concept of a system   47
Amplifiers and attenuators   48
Black boxes   49
Block diagrams   50
Amplifier distortion   50
Printed circuits   54

Chapter 5   Signals and data transmission   58

Information   58
Types of signals, analogue, digital and pulse   60
The sinewave   62
Bandwidth requirements   65
Modulation and demodulation   66
The binary system   67
Digital codes   69
Transmission   71
Different transmission systems — telegraphy, telephony, facsimile, radio and TV   71
Transmission errors   72

Chapter 6   Power supplies and amplifiers   78

Rectification using thermionic diodes   82
The pn junction and its use in rectifier circuits   86
Halfwave, full wave and bridge rectification   87
Batteries, primary and secondary cells   88
Charging and discharging of secondary cells   92

Chapter 7   Components   100

Passive and active components   100
Resistors, capacitors and inductors   103
Size, rating and manufacture   104
Symbols and coding   106
Connectors, valves and semiconductor devices   120
The transistor — bipolar and FET symbols   122

Logic gates   123
Use of logic circuits   125
Symbols   127

Chapter 8   Measurements and instruments   132

Pointer instruments, moving coil, moving iron,
   electrodynamic   133
Ammeters and voltmeters   136
Ohmmeters   139
The multimeter and its use in both d.c. and
   a.c. circuits   141
The digital display meter   144
Instrument errors   144
Frequency response measurements   146

Chapter 9   Reliability   153

Reliability philosophy   153
Concept of failure   154
M.T.T.F. and M.T.B.F. of a system   154
Series systems and the effect on failure rate   156
The 'bath-tub' curve   157
Preventive maintenance   159
Index   161

Chapter 1

The start and expansion of electronics

Introduction

An author of a textbook on 'electronics' is faced initially with the difficult problem of deciding where to begin. It is of course true to say that the start of any book poses its own special difficulties but 'electronics' is a subject where there are many possible beginnings, none of which is ideal.

One could start for example by explaining what an electron is but this is not very satisfactory because we cannot describe an electron but merely say what effects it can bring about. It would be similar to a situation in which a doctor could not describe a disease but only say what symptoms it produced.

An alternative beginning might be to delve back into history and talk about the derivation of the term 'electron'. The word is derived directly from the Ancient Greek word ελετρον (*electron*), meaning amber. Such a start is also not very fruitful and perhaps a word of explanation of how such an unlikely word became associated with 'electronics' would not be out of place. The Ancient Greeks around 600 B.C. had observed how the material amber when rubbed with fur had the property of attracting very light objects such as feathers or hair. This was very much later (in the eighteenth century A.D.) attributed to an electrostatic phenomenon but the term 'amber' became associated with this particular property and thus provided the root word for both 'electricity' and 'electronics'.

2 Early discoveries

This is obviously going back too far in history. Electronics is a twentieth-century science having its beginnings somewhere within the last 100 years. It was not however a science discovered by any one person or even groups of persons but came naturally out of a study of electricity, which itself was only just becoming established in the latter half of the nineteenth century.

There were many pioneers involved in the development of this new science of electronics — some were physicists, some were mathematicians (James Clerk Maxwell was an outstanding example) and some were really observant experimentalists who contributed to the growth of electronics, often unwittingly. It is difficult therefore to be precise and give the date when electronics became an identifiable science in its own right but if any attempt is to be made one can perhaps hazard a guess and suggest that the 'Electronics Era' began soon after 1875. About this time a number of research workers, notably in the UK, in the USA, and in Germany, had been aware of certain electronic phenomena but had not been able to find a suitable explanation for them. Crookes in the UK had discovered 'cathode rays' in 1879 and Thomas Edison in the USA had noticed a peculiar and unexplained effect in electric filament lamps (which turned out to be the emission of electrons by heat) around 1883.

It was Sir J. J. Thomson in a Royal Institution lecture in London on 30 April 1897 who first announced the existence of electrons. He did not at the time use the term electrons but referred to them as 'corpuscles'. He did however comment on their extremely small size. The actual words used in the conclusion to his historic lecture were:

'These numbers [referring to certain measurements he had made] seem to favour the hypothesis that the carriers of the charge are smaller than hydrogen atoms.'

This was an outstanding discovery but relatively few persons then understood this statement and even fewer appreciated the significance of Thomson's words. Seven years were to elapse after this quite sensational announcement before a practical application of the electron appeared.

Even today it is perhaps necessary to explain why this lecture was a milestone in the growth of electronics. The essential part of Thomson's conclusion was that electric current was not a steady fluid flow like liquid along a tube, as had been generally believed up to then, but the movement of extremely small (and separate) charged particles. The particles (actually electrons), while possessing individually an extremely small charge, existed in such large numbers that collectively they constituted a large movement of electric charge. The result was apparently a flow of electric current.

Unfortunately (or so it seemed) electric current was up to then considered to flow from the positive plate of a battery through the external circuit to the negative plate, which was in the opposite direction to the electron flow. Rather than change an agreed convention scientists decided to overcome this difficulty by ascribing a negative charge to the electron. Thus electron movement from right to left is exactly the same as conventional current flow from left to right (Fig. 1.1). Today with semiconductors we deal with both positive and negative moving charges and there is no real advantage in considering electric current as negative charges moving in the opposite direction or positive charges moving in the same direction (Fig. 1.2).

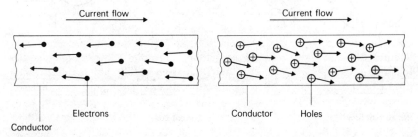

Fig. 1.1 Electron flow

Fig. 1.2 Movement of positive charges

The first electronic valve

In 1904 Sir Ambrose Fleming in the UK patented the first electronic device — a thermionic valve. It was developed more directly from the 'Edison effect' (already touched on) than from Thomson's discovery, in spite of the fact that its principle of operation depended upon the existence of electrons. The Edison effect came about briefly as follows and serves to illustrate how persons who were not 'scientists' in the accepted definition of the word contributed to the growth of 'electronics'.

Edison, involved in the manufacture of electric-filament lamps, noticed that after some time in use the glass of these lamps became blackened, which not only reduced the useful light output but also seemed in some way to have an adverse effect on the life of the lamp.

He introduced a plate or electrode into the lamp in an attempt to discover the cause of the blackening and noticed that a small electric current flowed from the filament to this electrode when a sensitive meter was connected between it and the positive end of the heated filament. The current ceased when the meter was connected between the electrode and the negative side of the filament (Fig. 1.3).

This effect can today be easily explained by the fact that free electrons are produced by the heated filament and they 'jump' the gap between the filament and the plate when a positive potential is applied to it, causing a flow of current. In 1883 this simple explanation was far from obvious.

Fleming's patent simply used the Edison effect in what was essentially a thermionic diode (a two-electrode electronic valve). Electrons could move inside the valve only from the heated filament to the plate electrode (the anode) and not in the opposite direction. The device had the property of a rectifier, allowing current to flow through it in only one way — hence the term 'valve' (Fig. 1.4).

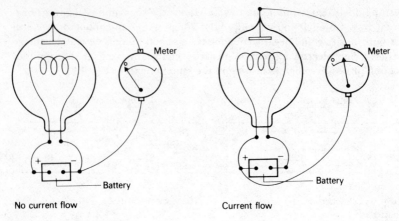

Fig. 1.3 Edison effect

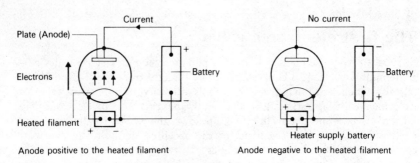

Fig. 1.4 Action of the diode valve

The discovery of the triode valve

Rather more than two years later Lee de Forest in the USA added a mesh of wires between the filament and the anode of Fleming's diode and showed how varying the potential on this mesh (the grid) could influence the flow of electrons through the valve (Fig. 1.5).

This flow could be cut off completely if a sufficiently large negative voltage was applied to the grid, since the negative charge on the grid repelled the negatively charged electrons from the heated filament or cathode. Small variations in grid voltage about a steady negative value (a bias voltage) caused a significant variation in the electron flow inside the valve and hence the current in the external circuit. Passing this current through a resistor produced a voltage which could be larger than the variation in the grid voltage. This was in many respects a considerable advancement on the Fleming valve as a thermionic device had now been produced which was capable of amplification, i.e. magnifying a voltage or current.

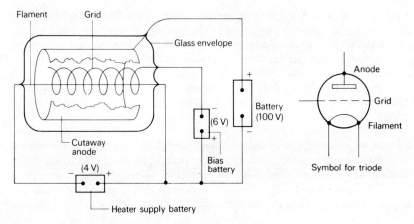

Fig. 1.5 The triode valve

The first three-electrode valves (triodes) were crude in construction and unreliable in operation but developments over the next few years steadily improved their performance. It was found that tungsten when heated produced a copious supply of 'free' or 'emitted' electrons and the electron movement in the valve became less impeded if all the air inside the glass envelope was removed.

The high-vacuum valve (in diminishing use today) was not, however, produced in any quantity until around 1912. At the present time (1977) thermionic devices have largely given way to transistors although in a few high-power and high-frequency situations the valve is still superior.

The start of the semiconductor era

Semiconductors have been with us many years in the form of rectifiers. During the Second World War copper-oxide rectifiers were in fairly extensive use, giving way to selenium rectifiers and finally the silicon rectifier. The first semiconductor device was the point-contact rectifier and appeared in 1915. It became very popular shortly after the end of the First World War with the establishment of broadcasting and audio entertainment. It consisted of a small piece of the material galena, a compound of lead and sulphur in crystalline form. Certain areas of this crystal had the property of allowing current to flow through them in one direction only. By searching with a fine wire these areas could be located. The fine wire carried the nickname of 'cat's whisker' and such rectifiers were extensively used in the early domestic radio receivers. Needless to say some patience had to be exercised by the user of the cat's whisker rectifier before satisfactory performance was achieved. It was a matter of luck in locating the right spot.

The production of the three-electrode transistor did not come about until 1948 — at least not as a commercial proposition — although on paper it had existed a quarter of a century earlier. This time lapse of 25 years between its discovery on paper and its actual production was due to two main reasons: (i) the incomplete understanding of the phenomenon of semiconduction; and (ii) the inability to produce semiconductor material of sufficient purity to enable the devices to function.

The first transistors were in fact point-contact devices and used two fine wires quite close together to produce the characteristics required of an amplifying device.

These gave way relatively quickly to the junction transistor — now referred to as the bipolar transistor. These in turn are gradually being replaced by field-effect transistors (F.E.T.s). It is strange to relate that these latest devices have characteristics which are very much akin to the thermionic valve — so much so that we use symbols to denote F.E.T. constants which are identical to the valve constants. But more about this later.

The interwar years

The First World War did not influence the growth of the electronics industry to any great extent and only the Royal Navy employed electronics in its day to day activities. In fact up to 1939 the expansion of the science was influenced as much by the entertainment world as anything. The establishment of the British Broadcasting Company and its London transmitter with its call sign '2LO' in 1922 at Savoy Place (the headquarters incidentally of the Institution of Electrical Engineers) produced a public demand for cheap and reliable radio receivers and in consequence cheap and reliable electronic components.

This resulted in the establishment of the radio valve industry with such manufacturers as Mazda, Osram, Brimar, etc. with the multifarious types of valves and valve sockets (4-pin, 5-pin, 7-pin, 8-pin (octal and 'loctal') 9-pin and even side-sockets).

The parallel development of the radio industry took place with its once household names (some still with us) such as H.M.V., Marconi, Ekco (E. K. Cole), Philco, McMichael, Scott-Taggart, Pye, Murphy, etc. Soon radio was no longer a mystery and by about 1935 very few homes did not possess (or at least did not have access to) a radio receiver.

The industry was given a further boost in 1936 when the world's first regular high-definition television broadcasts began from Alexandra Palace in North London. The essentially mechanical T.V. system devised by Baird had been in existence some few years but was quickly shown to be vastly inferior to the E.M.I. electronic high-definition system. Television sets employing the electronic system in those days were very expensive and possessed only small screens (25 cm (10 inch) or even less!). Nevertheless, electronics had clearly made its impact in the home and on the way of living.

Developments in the field of electronics were also occurring elsewhere. Navigational aids were making an appearance on seagoing vessels. Most of the bigger vessels already had their own radio transmitting and receiving sets largely using the Morse code to convey information, but radio direction beacons were beginning to appear, making ocean and air travel less hazardous. People like Sir Robert Watson-Watt were starting to experiment with radio reflections from moving objects as a means of locating these by radio. The forerunner of radar was in fact called radiolocation. Others began using radio to control remote objects such as pilotless aircraft. As the war clouds loomed in 1939 it was fairly obvious that electronics (a word incidentally still to be invented) was going to play a very important part in any hostilities.

The Second World War and its influence on the growth of electronics

When the Second World War broke out in September 1939, one for the first actions taken by the UK authorities was to stop all T.V. transmissions. This was because the relatively short waves employed by T.V. could have provided a 'homing' beacon for enemy aircraft. Television broadcasting ended on 1 September — two days before Great Britain entered the war. The 'radiolocation' towers which had appeared at many coastal points a year or two earlier were already manned day and night keeping watch for enemy aircraft. With a surveillance range approaching 200 km (120 miles) and the top speed of the bomber then less than 90 m/s (200 miles/h) the R.A.F. fighters could be given ample warning to be in a position to attack enemy aircraft. It is probably true to say that the Battle of Britain would have been lost without the help of radar. Another area of hostilities — the North Atlantic — was also influenced by electronics.

The enormous shipping losses caused by submarine warfare, reaching a peak of 600 000 tons per month in June 1942, were reduced mainly by the use of two electronic inventions. The first was the use of airborne radar known as A.S.V. (air surface vessel). This enabled submarines on the surface to be located by aircraft during night time as well as by day. The second was the invention of a new type of valve — the magnetron, capable of producing short high-frequency pulses of considerable power. Because of the very high operating frequency, radio transmissions from aircraft carrying this device remained undetected by the German submarines.

These and many more electronic devices employed by both sides in the war needed very large investment in both manpower and resources and resulted in a significant increase in the size of the electronics industry. It also provided a number of trained people in the armed services and in industry whose interest in the subject became aroused by their contact with it.

The post-war period

Britain was rather slow in making capital out of the war-time investment in electronics but, perhaps because of this, was better able to accept the change which the transistor brought about a few years later. The junction type transistor re-invented by Shockley and colleagues Bardeen and Brattain at the

Bell Telephone Laboratories in the USA in 1948 soon became a commercial proposition. The transistor challenged the thermionic valve almost from the beginning. It was cheaper, more rugged and had with careful use almost unlimited life. The fact that, unlike the valve, no power was needed for heating any filament to supply electrons meant that it was also more efficient. It also operated at a much lower voltage than the valve.

The result has been a revolution in the electronics industry which is still continuing, with components steadily improving in performance yet decreasing in both size and cost.

The computing industry which had started in this country soon after the war mainly in research departments in universities and government establishments initially employed thermionic valves which required enormous amounts of power and space. As an example an early electronic computer manufactured in the USA and given the name ENIAC (Electronic Numerical Integrator and Automatic Calculator) used 18 000 valves, 3000 indicating lamps, 5000 switches and consumed 150 kW of electricity. It is salutory to think that in a space of some thirty years the work of this computer can now be largely undertaken by certain programmable pocket calculators.

The revolution brought about by the introduction of the transistor generated a series of further changes, many of which are continuing today. Probably the most significant advancement in transistor manufacture was the change from germanium to silicon as the commonplace semiconductor material and the introduction of the planar technique in production. This has resulted in the microcircuit and has provided us with the ability of packing large numbers of electronic devices into a very small space and at a surprisingly cheap price. We have now as a consequence not only the pocket calculator but the electronic wristwatch. We also have small units which can be implanted in the human body to assist malfunctioning organs – the heart pacemaker is an example.

It is not impossible that eventually some form of sight may be available to blind persons by electronic means.

Future developments

We have certainly not seen the last of the developments in the field of electronics and the electro-biological area would seem to offer the most exciting prospects for the future. At present electronics feature in nearly all forms of activity – the automatic take-off, navigation and landing of aircraft – the ability of ships to 'see' by radar even in fog – the automatic control of manufacturing processes in industry – the ability to communicate via satellites to any part in the world – the ability to scan deep into space with the radio telescope – the capability of carrying out very complex calculations at high speed – and, perhaps a little disturbing, the invention of the laser. This device is able to produce an extremely narrow beam of light with the capability of enormous pulses of power of short duration. Such a device can pierce metal and provides unfortunately the basis of a lethal weapon.

Early researchers and workers in telegraphy

The expansion of the scence of electronics was considerably influenced by the coincident expansion of telecommunications. Indeed the two were inextricably mixed together and each depended upon the other.

The first large-scale organised attempts to 'communicate' at a distance depended neither upon electronics nor electricity but used a chain of semaphore stations, adjacent stations being within visual distance of each other.

Although transmission over long distances took time, it was certainly far quicker than sending short messages via a man on horseback and had the further advantage of being able to span difficult country (rivers or steep valleys) just as easily as flat terrain. Napoleon Bonaparte used this method extensively and his military successes were certainly influenced by his relatively rapid communication system.

In the early 1800s the possibility of using electricity as a means of conveying signals over large distances began to assume reality. In 1809 a German, von Soemmering, produced a form of telegraphic system whereby communication was established over a few metres using electricity as the communication 'agent'. The first working magnetic telegraph (on which the whole science of telegraphy was based) was demonstrated by another German, Paul von Schilling-Cannstadt, in 1835 and from this invention came the electric telegraph of Wheatstone and Cooke in the UK in 1837. It was quickly taken up by the Great Western Railway as a means of conveying messages from station to station and provided the first public communication service in this country. It hit the headlines in 1845 when a murderer was apprehended in London following a telegraphic message sent from Slough to Paddington.

About the same time as Wheatstone and Cooke's inaugural telegraph, Samuel Morse in the USA demonstrated a successful telegraph system using a code whereby each letter of the alphabet was represented by a system of 'dots' and 'dashes' which enabled letters (and hence messages) to be conveyed by a sequence of switching. Although Morse was not the originator of this form of telegraphic switching the famous 'Morse code' is with us still.

There was not as much interest shown in the telegraph system initially as Morse had hoped and it was some time before people saw the tremendous advantage of being able to send messages quickly over relatively long distances. As in the UK it was the railroads which adopted the telegraph in the first instance. But there were technical difficulties in sending telegraphic signals over long distances. Losses in the telegraph lines caused the signal to attenuate, i.e. lose in strength, and the only way to boost the signal seemed to lie in the use of larger signals at the sending end, which meant employing higher voltages.

The successful spanning of the Atlantic in 1866 by a telegraph cable required for its operation large sending-end voltages and sensitive detectors at the receiving end. It was unfortunate that the lack of understanding of the transmission problems and the non-availability of adequate insulating materials caused the early submarine cables to fail soon after installation.

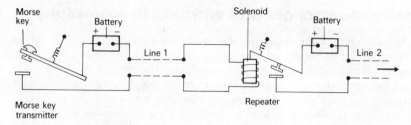

Fig. 1.6   The telegraph repeater

The attenuation problem on land was solved by the invention of the 'repeater' (Fig. 1.6) which virtually regenerated the Morse signals at their original power and (almost literally) boosted the signal. But sending messages by means of tapping out a series of dots and dashes required the services of a skilled telegraphic operator and in consequence telegrams were expensive. Furthermore, it was still a relatively slow and laborious process and hardly conducive to two-way conversation.

From telegraphy to telephony

The invention of the telephone by Alexander Graham Bell, a Scottish speech therapist, in the USA in 1875 was a significant advancement in telecommunications and quickly resulted in an urban network of wires strung on poles in nearly every large city in the USA and very soon afterwards in Europe. Telephone exchanges (manually operated) enabled 'caller' and 'called' to be connected to each other. It was even possible in Paris to listen to the opera in one's home by means of the telephone.

There was however the same snag as with the telegraph, namely that over long distances the signals became attenuated and the repeater employed in the telegraph could not be used with telephone messages. The telegraph repeater employed a straight switching process and, being electromechanical in operation, was relatively slow. Speech signals were vastly more complicated and needed an inherently higher signalling speed (voice frequencies rather than the frequency at which the Morse code could be tapped out).

It was not until December 1906 with the invention of the triode by de Forest that the means existed of amplifying the speech signals. It was at this point that the interaction of 'electronics' and 'telecommunications' really began in earnest.

Wireless telegraphy (radio)

The growth of the telegraph and the telephone was always limited by the fact that the transmitter and receiver had to be joined by conducting wires. A number of researchers had been trying to transmit signals without conducting wires. Heinrich Hertz in Germany had succeeded in transmitting 'radio' waves across a room in 1888 and James Clerk Maxwell had demonstrated mathematically the possibility of radio transmission, but 'wireless telegraphy' became a reality around the year 1900.

Guglielmo Marconi, of mixed Irish/Scottish and Italian parents, had shown an interest in telecommunications at quite an early age and although having had no formal scientific training soon demonstrated an uncanny understanding of the problems involved. Marconi came to London in 1896 and was soon in touch with such persons as A. Campbell Swinton (responsible for the design on paper of a workable electronic television system in 1912) and William Preece, Engineer-in-Chief of the Post Office. Marconi received advice and encouragement so that by 1901 he was able to demonstrate the transmission of radio signals across the Atlantic and dispelled the existing theory that radio waves could travel only in straight lines. His early experiments used the Morse code for the transmission of radio signals and hence earned the name of 'wireless telegraphy'. Wireless or radio telephony did not come about until well after 1906 when once again the thermionic triode valve became so essential to the further development of radio.

The First World War

The outbreak of hostilities in 1914 and the ensuing years of war had remarkably little effect on the advancement of either electronics or telecommunications. There was a steady improvement in quality and reliability of the electronic apparatus but no significant invention or 'breakthrough' occurred. The Royal Navy was the only branch of the armed services to make significant use of radio telegraphy and this was due largely to the fact that there was no other means of long distance communication with vessels at sea.

A directional aerial radio system was employed in 1916 which enabled movements of the German fleet to be detected before the Battle of Jutland but other than this radio telegraphy was still the only widely used sector of the emergent science.

As has already been mentioned it was the establishment of widespread entertainment by radio and television which had by far the greatest impact on the growth of the electronics industry and telecommunications.

Professional bodies and their influence on electronics

The many pioneers in the fields of electronics and telecommunications were scattered world-wide and their efforts were coordinated only when they began to communicate with each other. Initially much of the communication was by personal contact or letters but the more significant exchanges occurred in lectures and ensuing discussions. Two institutions in England where such activities could take place were already well established — the Royal Institution and the Royal Society — and significant publications of scientific progress in both electronics and telecommunications came from these bodies.

The first established UK organisation to be specifically interested in telecommunications was the Society of Telegraph Engineers set up in 1871, although earlier attempts at setting up a similar body had proved abortive. This society was the forerunner of the present Institution of Electrical Engineers and as its name suggests was initially interested mainly if not solely with telegraphy.

The Institution grew in size and broadened its interest and in fact its central activity moved more towards the 'power' side of electrical engineering in its first fifty years of existence. Even so a number of significant papers were delivered which demonstrated its original interest in telecommunications. One such paper was that by Sir Oliver Lodge in 1898 entitled *Improvements in Magnetic Space Telegraphy*, a foretaste of radio.

It was, however, not until 1919 that the Institution of Electrical Engineers showed publicly its interest in radio when the 'Wireless Section' was set up.

The Institution of Electronic and Radio Engineers (originally the British Institution of Radio Engineers) was not founded until 1925 and had therefore no effect whatsoever on the early work in electronics. It has nevertheless contributed to its development significantly since that date.

Both these institutions (and others) provided the necessary focus for the exchange of ideas, theories, and viewpoints in the field of electronics and telecommunications. Today the hallmark of the professional electronic engineer is the election to corporate membership of one or other (or both) of these institutions.

A calendar showing the main events in the development of electronics and telecommunications up to 1950 is given in Table 1.1.

Table 1.1

Year	Event
1816	Design of a viable telegraph system by Sir Francis Ronalds in London.
1835	First demonstration of the magnetic telegraph in Bonn by Paul von Schilling-Cannstadt.
1837	First working magnetic telegraph installed in UK by Wheatstone and Cooke.
1851	English Channel telegraph cable successfully completed and telegraph link with France established.
1857	First and unsuccessful attempt to span the Atlantic by telegraph cable.
1865	Telegraphic link from England to India established.
1866	First successful transmission of telegraphic signals across the Atlantic.

Table 1.1 cont

Year	Event
1871	I.E.E. founded as the Society of Telegraph Engineers.
1875	The invention of the telephone by Alexander Graham Bell.
1887/88	Heinrich Hertz demonstrates radio waves.
1897	Sir J. J. Thomson's historic lecture on the existence of electrons.
1901	Marconi demonstrates successful spanning of Atlantic with radio telegraphy.
1904	Invention of the diode by Sir Ambrose Fleming.
1906/07	Lee de Forest's invention of the triode.
1912	Titanic disaster — first use of the SOS distress call using 'wireless'.
1914/18	First World War — development of radio telegraphy mainly by the Royal Navy. Battle of Jutland (1916) was largely the result of detection of the movement of the German fleet by radio.
1920	First flight of a civil airliner fitted with radio transmission.
1922	B.B.C. began regular entertainment radio broadcasts.
1925	British Institution of Radio Engineers founded.
1926	Experimental T.V. station opened using the Baird system.
1936	First regular transmissions of 'high-definition' television from Alexandra Palace.
1939/45	Second World War — radar and all its associated inventions in high-power, high-frequency pulse transmission.
1946	First large-scale electronic computer (ENIAC) commissioned.
1948	Junction transistor appeared as a commercial product.

Objective tests on Chapter 1

In the following statements (a)–(m) which of the alternatives is correct? Write down the number of the alternative you select.

(a) The discovery of the electron is attributed to
1. Thomson
2. Edison
3. the Ancient Greeks
4. Fleming

(b) The first electronic device was patented by
1. Marconi
2. Fleming
3. Lee de Forest
4. Hertz

(c) A semiconductor device was first employed
1. before the First World War

2. during the First World War
3. between the First and Second World Wars
4. during the Second World War

(d) The triode valve was first in use
1. before 1880
2. between 1880 and 1900
3. between 1900 and 1910
4. after 1910

(e) The most significant advance of the triode valve over the diode valve was
1. its ability to rectify
2. its ability to amplify
3. its ability to operate at high frequency
4. the fact that it was a thermionic device

(f) The most significant advantage of the transistor over the triode valve was
1. its ability to amplify
2. its ability to handle high powers
3. the absence of a heater supply
4. its ability to rectify

(g) The growth of the electronics industry up to the beginning of the Second World War was largely due to
1. the impact of the First World War
2. Marconi
3. Baird
4. the impact of broadcasting

(h) Public radio broadcasting on a significant scale began in the UK
1. before 1918
2. between 1918 and 1921
3. between 1921 and 1924
4. after 1924

(i) Regular T.V. high-definition broadcasting began in the UK
1. in 1922
2. between 1923 and 1935
3. in 1936
4. after 1936

(j) Telegraphy was first used as a military aid. The military leader who first used this extensively and successfully was
1. Napoleon
2. Wellington
3. Alexander the Great
4. Field Marshall Montgomery

(k) Which of these statements is correct?
1. The transistor was used extensively during the Second World War

2. The transistor was first produced in 1948 but had been invented on paper about a quarter of a century earlier
3. The transistor has now replaced the thermionic valve so that today (1977) the valve-manufacturing industry is virtually dead
4. Today (1977) transistors have replaced the thermionic valve in all electronic applications.

(l) A typical pocket electronic calculator contains the equivalent of
1. up to 10 transistors
2. up to 100 transistors
3. up to 1000 transistors
4. more than 1000 transistors

(m) The main material used today in the manufacture of semiconductor devices is
1. silica
2. silicon
3. silicone
4. germanium

Essay-type questions

1. Write a short essay (about 600 words) on ONE of the following subjects:
   (a) Communications — use and abuse
   (b) Communications in the year 2000
   (c) The need for good communications
   (d) The impact of T.V. on society

2. If a national catastrophe occurred and normal communication systems were destroyed, suggest the order in which you would attempt to restore the following, giving brief reasons for your choice:
   (a) the television service
   (b) the telephone system
   (c) radio broadcasting
   (d) the postal service
   (e) the rail system
   (f) the road system

3. The need for cheap electronic components came about with the demand for cheap but reliable radio receivers. Which of the following do you think was largely responsible for the production of low-cost components? Substantiate your answers with brief reasons.
   (a) Standardisation of electronic components
   (b) Competition between major electronic manufacturers
   (c) The advancement of technology
   (d) Improvement in the education of electronic technicians

4. Since the Second World War there has been a growing demand for further education in the field of electronics. To what extent is this due to the demands of:
   (a) school leavers?
   (b) teachers in schools?
   (c) industry?
   (d) the Colleges of Further Education?
   Write short notes on each of these alternatives.

Chapter 2

The electronics industry

Early days

The start of the electronics industry obviously followed the invention of the thermionic valve although before this there existed an extensive industry based on the fields of telegraphy and telephony. Cable manufacturers were well established during the mid 1800s in the UK, having their factories situated mainly along the Thames estuary, and in fact some are still to be found in the Woolwich area. Following the invention of the telephone a further industry built up but it was very much influenced by the Post Office, which quickly obtained an almost total monopoly of all communication systems in this country. This was in sharp contrast to the situation in the USA, where private enterprise was allowed to flourish — often with extensive argument as to who invented what and who had exclusive rights to manufacture certain components and apparatus.

Although the Post Office in the UK controlled the use of all communication systems and issued transmitting licences rather grudgingly, it did not influence the design of radio receivers. The production of the components making up the radio receivers and the manufacture of the receivers themselves was very much left to individual companies and private enterprise. This had the advantage of competition with the corresponding competitive prices, bringing radio within the financial limits of many families. It had also the disadvantage of non-standardisation — at least in the early days — so that valves, for example, could not always be readily interchanged from one manufacturer's set to another.

The Marconi Company

One of the first (if not the first) companies in the world to be concerned with radio transmission and reception was The Wireless Telegraph and Signal Company Limited, registered in 1897. This was set up by Guglielmo Marconi in Chelmsford, Essex. Three years later its name was changed to 'Marconi's Wireless Telegraph Company Limited'. Marconi had come to the UK from Italy (where he was born) mainly for two reasons. Firstly, no one whom he approached in Italy seemed to consider that his inventions had any future. Secondly, he had one or two influential friends in Britain, among them Henry Jameson-Davis (actually Marconi's cousin), who managed to persuade William Preece, then Engineer-in-Chief of the British Post Office, to see a demonstration of Marconi's radio transmission. From these initial overtures came the Marconi radio factory. The staff for this venture — all initially quite unaware of the new technology — were recruited from the technical schools of the era.

The newness of radio was such that orders for equipment came in very slowly indeed and Marconi and his staff were often more concerned with advertising radio to the uninitiated than actually producing radio sets. Marconi concentrated initially on radio communication from ship to shore. By 1902, twenty-five shore radio stations had been set up and about seventy vessels were equipped with the Marconi transmitting and receiving sets. The apparatus was rented rather than sold to the users.

The Wireless Telegraphy Act of 1904 gave the Post Office complete monopoly of radio communication in the UK and in consequence Marconi's shore stations had to be given licences to enable them to continue to operate. The Post Office authorities actually forced Marconi to sell the stations to them in 1909.

Marconi, who in 1901 had demonstrated that trans-Atlantic radio communication was possible, set up a regular trans-Atlantic service in 1907. But even with all these innovations the Marconi Company did not prosper and was forced to economise. Some of the senior staff did not receive any wages and the work force in the factory was reduced by sacking 150 workers. Marconi had to invest much of his personal wealth to keep the company in business. It was not until 1910, when a Godfrey Isaacs was appointed as managing director to the firm, that fortunes changed and the company paid its first dividend to the shareholders. The company survived a scandal in 1912 after it had negotiated a large contract with the government. It appeared that certain ministers with prior knowledge of the contract had invested heavily in Marconi shares and although later this was proved largely untrue it delayed the contract by a year.

While Marconi was undoubtedly a leader in the field of radio he did not anticipate the growth of the thermionic valve industry. Although he had secured patent rights on Fleming's diode he failed to appreciate the significance of the invention of the triode, and had it not been for the foresight of one of his senior engineers (H. J. Round) the company would have lagged behind its competitors.

The war years (1914—18) did not unduly influence the growth of the electronics industry in general and the Marconi Company in particular, even with contracts placed by the Royal Navy for marine radio equipment. It was after the war when the real expansion occurred. Again it was Marconi himself who contributed to this growth when he set up a chain of transmitting and receiving stations abroad and offered a world-wide communication service. With the advent of broadcasting his organisation was well able to cope with the demands suddenly made for radio receiving sets by the general public.

His name was associated with many electronic and radio ventures — Marconiphone, the Marconi Valve Company, etc. and the word 'Marconigram' was freely used as depicting a message from or to a ship at sea.

The Company still exists at Chelmsford but is now part of the General Electric Company, which took it over in 1968.

Marconi himself died in 1937 after a series of heart attacks. His personal life was not an overwhelmingly happy one and he rather lost favour in British eyes when he supported the Italian dictator Mussolini. He must nevertheless be regarded as a pioneer of radio communication, an experimenter and an entrepreneur with vision.

Industry today — some philosophical views

The method of growth of the Marconi Company may have been copied by a number of electronics firms before the Second World War but the total investment cost of industry is now so large that no one individual or even a small group of individuals could possibly set up an organisation which could challenge the industrial giants of today. This is not to say there is no place for small firms, but these tend to be either part of a larger organisation or specialised production units aiming at small markets.

While a number of the older names still exist, such firms are seldom independent but are part of a group of companies. This does unfortunately create large monopolies where eventually there may be no competitors in a particular line of production, but it does also create the possibility of more efficient deployment of resources. In any case, although there may be only one producer of a product in the UK we still have competition from overseas and we have to sell our goods in other countries. There is also political pressure from the Labour government to bring firms together in a nationalised industry. We have today a National Coal Board, a nationalised electricity supply, a national steel industry, and so on.

There are arguments both for and against such organisations. Where exceptionally large financial investment is involved there is perhaps a stronger case for rationalisation, but where an industrial sector appears to be operating very successfully it seems to be a questionable move to step in and nationalise. One inevitable (or so it seems) result of large organisations is the setting up of a bureaucratic situation, i.e. one in which a great deal of unnecessary paper, form-filling, and office work are involved. Such activity is counter-productive and only pushes up the cost of a product. While lack of administration will lead quickly to chaos, too much can lead to frustration and stress situations. There is an optimum condition to aim for. The UK is unfortunately in economic difficulties (although many Western European countries are in somewhat similar circumstances). Making changes, however, is not as easy as it sounds as often there are so many different interests to take into account — the trade unions, the management, the financiers, the shareholders and the employees. If making a firm more efficient involves giving notice to 500 employees, industry might not wish to take such action. On the other hand, if a particular firm is rapidly losing money and the only way to prevent bankruptcy is to sack 500 employees then such action will have to be taken if only to save the jobs of perhaps 1500 others in the same organisation. The decision on who to retain and who to sack is sometimes extremely difficult.

Industrial management today is fraught with these problems and being a manager is a tough job with a great number of anxious moments and stressful situations. The popular concept of a successful manager as one who spends his time eating expensive lunches and playing golf most afternoons is a fallacy.

Equally well the concept of a factory worker as one earning fabulous sums of money for very little effort is quite wrong.

We can all be misled by biased views or press reports which are sensational but often only partially true. We must also always guard against judging the majority by the actions of a small minority. We must not be over-influenced by a noisy section of the community which, however well intentioned, represents the views of only a few.

The electronics industry today

There are in the UK only a few major manufacturing organisations entirely concerned with electronics as such, although a large number of firms use electronic equipment. Some of these giants are associated with firms overseas (e.g. Mullards and Philips of Holland) or part of an overseas organisation (e.g. Tektronix). One extensive user of electronic equipment is in fact the Post Office, the largest single employing body in this country (about 430 000 employees, although about 45 per cent of these are in the postal service and are not concerned with telecommunications). The total number of employees in the electronics manufacturing industry, which does not include users such as the Post Office, is around 443 000 persons (1974) and about 45 per cent of these are with one firm (General Electric Company).

The total value of manufactured electronic equipment in 1974 was £2409 million with exports accounting for £811 million.

The electronic components industry is centred on smaller firms (about 830 in number) and the deliveries of active components (transistors, valves, etc.) were valued at £162 million in 1974. Passive components (resistors, capacitors, etc.) accounted for a further £432 million in the same year.

Another large sector of the electronics industry is devoted to the entertainment field. The output of T.V. and radio sets in 1974 was 2·6 million T.V. sets, 1 million radios and 600 000 record players, amounting to a value of £432 million. The B.B.C. and I.B.A. are two further large organisations in this sector but are not manufacturers of equipment.

Computers and associated equipment had an output of about the same order in financial terms.

Radar and navigational aids were responsible for around £482 million.

It is obvious that the electronics and telecommunications industry is contributing significantly to the economy of the UK and is responsible directly for the employment of a very large number of people in a wide variety of jobs. Furthermore, a large amount of electronics is employed in sectors of industry not directly concerned with electronics, e.g. aerospace, car manufacture, etc.

The analysis of the electronics and associated industries is shown by Table 2.1.

Table 2.1

Manufacturing industry includes:

1. Electronic equipment manufacture.
2. Instruments and measuring equipment.
3. Telecommunication equipment and systems.
4. Component manufacture including integrated circuits.
5. Radio, T.V. and associated apparatus.
6. Computers and control systems.
7. Radar and navigation aids.

Users of electronic and telecommunication equipment include:

1. The Post Office.
2. The B.B.C. and I.B.A.
3. The aerospace industry.
4. Airlines (British Airways).
5. Shipping companies.
6. The armed forces.
7. Other non-electronic firms, e.g. the car industry.
8. Scientific and medical research organisations.
9. Local government, e.g. traffic control, police, highway maintenance, education.

To get some idea of the size of the electronics industry in relation to industry as a whole in the UK the following figures (1974) apply.

Total number in employment of all kinds:
24 000 000 including nearly 2 000 000 employers and self-employed.
1 500 000 in transport and communications.
1 600 000 in local and national government.
8 000 000 in manufacturing industry.
Of this last category there are around 92 000 firms of which:
> about 64 per cent employ less than 25 people and account for 70 per cent of the working population;
> about 32 per cent employ between 25 and 500 and account for 44 per cent of the working population;
> about 3 per cent employ between 500 and 2000 and account for 28 per cent of the working population;
> less than 1 per cent (409 firms) employ more than 2000 and account for 21 per cent of the working population.

i.e. about half the working population in the manufacturing industry are found in some 2800 firms which have more than 500 employees.

The structure of a single firm

Obviously the structure and organisation of industry will vary considerably from firm to firm and from sector to sector. With the small firm there is a tendency for the majority of persons employed in it to take on a wider variety of tasks. There is within one sector of industry however a remarkable similarity between firms of similar size. Two T.V. manufacturers, for example, each employing 1000 personnel will have a similar disposition of labour force. The organisation which is now described is not based on any particular firm but is fairly typical of the larger organisation (more than 500 employees) in the electronic equipment sector.

The first thing to consider is what the product consists of that the firm is making. Electronic equipment, e.g. measurement and test gear, requires active and passive components; steel or aluminium for the front panels and boxes in which to place the equipment; indicating meters; knobs, brackets, cable, sockets, plugs, etc.; insulating boards on which to place the components; and so on.

Unless the firm is producing its own integrated circuits it will most likely purchase active and passive components in bulk from other suppliers. The metal will be bought largely in sheet form of different thicknesses, but some small quantities of rod and tube might be needed for specially made components. It is also possible that some steel angle will be needed to make equipment racks. The firm will therefore be largely engaged in the production area in assembling, mounting, wiring and testing its product. There will need to be a stores (or a number of stores) in which to place all the components and material purchased.

The organisation will need to cut, bend, drill and shape the material received so there will have to be a section (the machine shop) to deal with this.

Many of the fabricated parts will need painting or plating to avoid corrosion so a paint shop and a plating shop will be required.

The parts will have to be put together – an assembly shop is needed.

The circuits will have to be wired so that a wiring shop is called for.

The equipment, once assembled and wired, will need to be tested and calibrated. There will have to be some check on the standards employed in the firm so a standards department is required.

Finally there is a 'finished goods' store where the final product is placed before dispatch to the customer. The goods will require packing before transit.

Behind all this activity will be a number of other bodies. Firstly, there is the administration dealing with correspondence to customers and suppliers of components, the invoicing (i.e. asking for payment) of the customers and the payment for goods received. This section (finance) will have to deal also with the payment of wages and the general assessment of the flow of cash to and from the firm.

There is a requirement to monitor the progress of the various jobs through the firm to ensure that the stores have an adequate supply of the right components and materials. This is production control. Then there are the design, research and development areas. The design section is concerned with the design of new equipment or the modification of existing products. The research section will be thinking of the future requirements, and in liaison with the development department will be planning the next generation of test gear that is to be marketed.

There are other vital areas in the firm – canteen workers who provide the meals; a social section looking after safety, first aid; training and personnel; etc. Overall it is the management which has to make decisions on how much of a particular product to make, when to change a design or discontinue a product line. The management is supported by its sales force and forward planning department, both of which must work closely with the finance office and the research and development group.

A firm as described is a complex organisation and each section is dependent on others. Remove one section and the organisation breaks down, although the absence of some sections will have a more rapid effect than others, e.g. remove the wiring section and production stops and only completed units awaiting test calibration and despatch continue. Close down the research and development department and no immediate effect occurs – production continues but with no new product planned the firm can no longer compete with similar firms and eventually sales decline and the firm is faced with bankruptcy.

It is unfortunate that when industry is forced to economise it is these areas which are cut with disastrous 'long-term' effects.

24   The organisation can be shown on suitable charts (Figs 2.1 and 2.2).

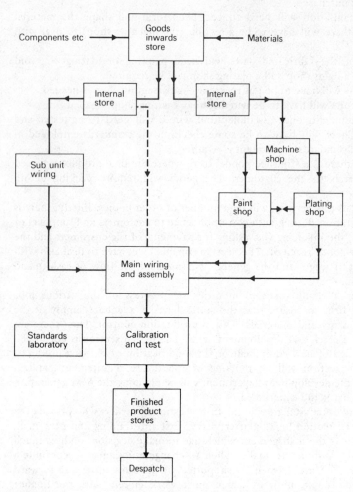

Fig. 2.1   Production unit

The management and administrative structure is highly interdependent and the senior manager of each of the main areas, finance, production, forward planning, sales and personnel, should meet regularly as the management unit under the general direction of a managing director or chairman.

The structure outlined is essentially one of hierarchy or 'order' and does sometimes suffer from being cumbersome and slow to respond to events. Other types of organisation have been tried but with no evidence of any improvement. The advantage of a system as described is that there are clearly defined lines of communication. It is often the persons within the system who are slow to respond rather than the system as a whole.

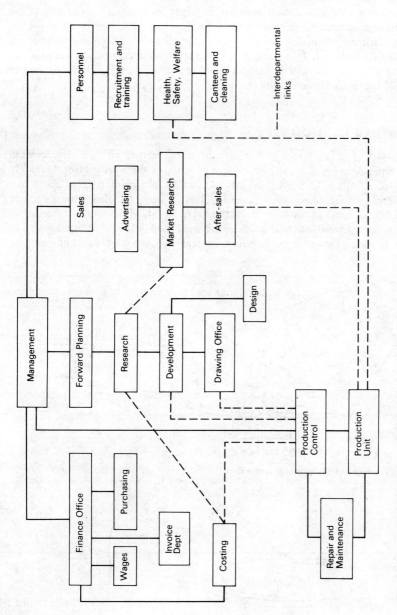

Fig. 2.2 Management and administration

Clearly, the areas of industry which are non-productive (in the sense of not actually producing some hardware) but offer a service are equally important but are organised on somewhat different lines. The Post Office and the British Broadcasting Corporation are two obvious examples. Each, however, have customers who are provided with a service which has to be acceptable in standard and cost. It is perhaps unfortunate that the former is a monopoly with no competition, although the telephone service is still a cheap method of rapid communication despite the recent price increase.

Safety in industry

Industry is by law forced to consider the safety factors within its organisation in order to prevent accidents. It must ensure that the workers are adequately protected from as many hazards as possible. No smoking is allowed, for example, where combustible gases or fluids are being handled. Gloves and eye shields as well as protective clothing are provided in areas dealing with corrosive materials such as acid. In the workshop, safety shoes are advised to prevent accidents to persons caused by walking over sharp edges or dropping

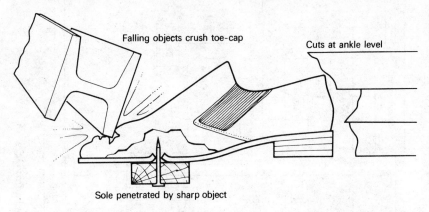

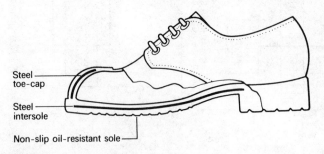

Fig. 2.3   Industrial safety shoe

heavy weights on one's toes (Fig. 2.3). But there are a number of points of potential danger which the individual must be aware of, because he may be largely responsible for an accident. Loose clothing, ties and long hair may be caught in rotating machinery with sometimes very serious injury or even death (Fig. 2.4). Guards are fitted to machines to minimise this type of accident but sometimes people remove these guards because they get in the way of a job. The electronics technician is less likely to meet these sorts of hazard because the environment in which he is working probably does not contain rotating machines. But he is working with hot soldering irons which can cause a nasty burn and an electrical supply which can kill. Long, trailing electrical leads should be avoided and the technician should work on an insulated floor which is free from damp or liquid and is slip-proof.

Nevertheless, however safety-conscious a person or a firm is, accidents will still happen. All accidents must be eventually reported to the firm's safety officer but in an emergency one has to know what to do immediately. In the case of electric shock the first thing to do is remove the person who has had the shock from the electric source. Switch off the supply if possible and drag the person away from the electric source by his clothing, ensuring

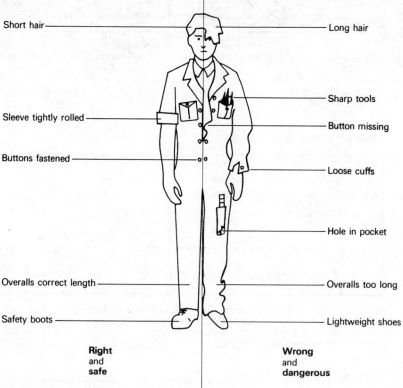

Fig. 2.4 Correct dress

Order of action

1 Switch off current
Do this immediately. If not possible do not waste time searching for the switch.

2 Secure release from contact
Safeguard yourself when removing casualty from contact. Stand on non-conducting material (rubber mat, DRY wood, DRY linoleum). Use rubber gloves, DRY clothing, a length of DRY rope or a length of DRY wood to pull or push the casualty away from the contact.

Method - mouth-to-nose

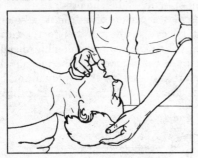

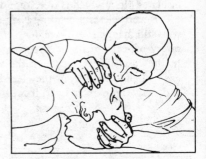

1. Lay casualty on back; if immediately possible, on a bench or table with a folded coat under shoulders to let head fall back. Kneel or stand by casualty's head. Press his head fully back with one hand and pull chin up with the other.

2. Breathe in deeply, bend down, lips apart and cover casualty's nose with your wide open mouth. Close casualty's mouth with your cheek or fingers of the hand holding the chin. Breathe out steadily into casualty's lungs. Watch his chest rise.

Method - mouth-to-mouth

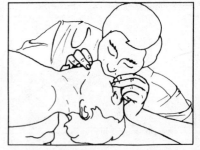

1. Lay casualty on back; if immediately possible, on a bench or table with a folded coat under shoulders to let head fall back. Kneel or stand by casualty's head. Press his head fully back with one hand and pull chin up with the other.

2. Breathe in deeply. Bend down, lips apart and cover casualty's mouth with your well open mouth. Pinch his nostrils with one hand. Breathe out steadily into casualty's lungs. Watch his chest rise.

Fig. 2.5 Treatment for electric shock

3 Start artificial respiration

If the casualty is not breathing artificial respiration is of extreme urgency. A few seconds delay can mean the difference between success or failure. Continue until the casualty is breathing satisfactorily or until a doctor tells you to stop.

4 Send for doctor and ambulance

Tell someone to send for a doctor and ambulance immediately and to say what has happened. Do not allow the casualty to exert himself by walking until he has been seen by a doctor. If burns are present, ask someone to cover them with a dry sterile dressing.

3. Turn your own head away. Breathe in again.

Repeat 10 to 12 times per minute.

If you have difficulty in blowing your breath into the casualty's lungs, press his head further back and pull chin further up. If you still have difficulty, check that his lips are slightly open and that the mouth is not blocked, for example, by dentures. If you still have difficulty, try the alternative method, mouth-to-mouth, or mouth-to-nose, as the case may be.

When casualty begins to breathe

Adjust movement to rhythm of casualty's breathing; breathe into casualty's lungs as his chest rises.

When casualty recovers

Keep casualty at rest. Remove on a stretcher. Watch closely, particularly for difficulty in breathing. Keep covered with blankets or other warm material.

3. Turn your own head away. Breathe in again.

Repeat 10 to 12 times per minute.

External cardiac massage

In a proportion of cases rendered unconscious by electric shock, the heart's action may be affected. The technique of external cardiac massage which might then be life-saving, cannot readily be taught on a notice. Therefore those people who are commonly working with electricity are encouraged to take practical instruction in cardiac resuscitation as taught by the major first-aid organisations.

Instructions prepared under medical advice.

that you do not sustain a shock yourself in this way. If the person is unconscious artificial respiration is needed and should be applied preferably by the mouth-to-mouth method. If the heart is not beating an external cardiac massage will be needed (Fig. 2.5). Usually there are people around who can provide help in an accident and if professional aid is immediately available (e.g. a qualified first-aid man) he should carry out the necessary procedure to the person who has had the accident.

Meanwhile you as perhaps an onlooker should have obtained other help either from inside the factory (if it is a large one with medical help available) or from outside by obtaining the appropriate authority (fire, police, ambulance). Do ensure, however, that you have a real emergency on your hands before summoning help from these outside organisations. A false alarm could easily prevent an ambulance answering a real emergency call.

The procedure in the case of an accident should be clearly understood by everyone and it is the responsibility of the factory management to ensure that everyone has explicit instructions as to what to do in these circumstances. A fire drill, for example, should be held periodically and all emergency exits must be kept clear of obstruction. It is everyone's responsibility to see that the safety measures and regulations provided are adhered to and any contravention of these regulations is drawn to the attention of the senior personnel as soon as it is noticed. Safety is everyone's responsibility.

Exercises

1. From the information given in this chapter draw pie-charts (i.e. circles with different size sectors) for:
   (a) the total working population of 24 million in terms of the type of industry in which they are employed;
   (b) the distribution in terms of money from the different sectors of the electronics manufacturing company;
   (c) the distribution of the working population in the manufacturing industry in terms of size of firms in which it is employed;
   (d) the number of firms in the manufacturing sector in terms of size.
   (The last two pie-charts should be arranged for comparison.)

2. Draw an organisational chart of the firm in which you are employed. If it is a large organisation your training officer or personnel officer will be able to supply you with information.

3. To what extent do you think that a large firm has advantages and disadvantages compared with the small firm? Give your answer in the form of a short essay of less than 300 words.

4. If an accident involving an electric shock occurred to one of your colleagues, state briefly what you would do.
   (N.B. The information needed to answer this question will have to be found from the firm in which you work. It is an important question

which needs answering, if only for your own safety. An accident might happen to you.)
Who would you inform?
    The T.U.C.?
    The shop steward?
    The managing director?
    Your immediate superior?
    The police?
    The personnel manager?
    The training officer?
    The welfare centre?
    Others? Please specify with reasons.

Chapter 3

The electric circuit

Ohm's law

The study of any branch of electrical science is based on a number of fundamental laws. One of the most important is Ohm's law and this must be clearly understood before any detailed study of electronics can take place.

Ohm's law is concerned with three electric quantities – voltage (V), current (I) and resistance (R).

In an electrical circuit we must have a source of electrical energy and in the simplest case this can be a single cell. A source of electrical energy has a certain 'electrical pressure' and the 'pressure difference' or 'potential difference' measured across the plates of the cell is recorded in **volts**. The potential difference (p.d.) of a typical flash-light cell is about 1·5 volts.

If such a cell is placed in series with a switch and a torch 'bulb' (Fig. 3.1) and the switch is closed, we then have a flow of electric current from the positive terminal to the negative terminal and its effect can be seen as the bulb lights up. The current is measured in **amperes** and its magnitude depends upon: (i) the pressure (p.d.) producing it; and (ii) the construction of the bulb which 'resists' the flow.

Ohm's law merely states that the current flowing in such a circuit is related to the voltage producing the current and the 'resistance' of the circuit.

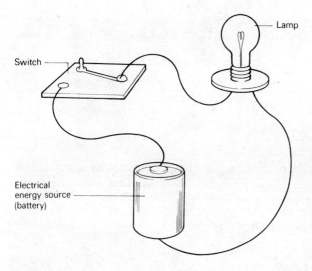

Fig. 3.1 Simple electric circuit

If the p.d. is in volts, the current in amperes, the resistance is measured in **ohms**; then mathematically

$$I \text{ (current in amperes)} = \frac{V \text{ (p.d. in volts)}}{R \text{ (resistance in ohms)}}$$

This is in effect a statement of Ohm's law.

The abbreviation for amperes is A, the abbreviation for volts is V and the abbreviation for ohms is Ω (the Greek capital letter *omega*).

Thus if a voltage of 25 V is applied across a resistance of 10 Ω the current flow is

$$I = \frac{25 \text{ V}}{10 \text{ Ω}} = 2 \cdot 5 \text{ A}$$

The form of Ohm's law just expressed, namely

$I = \frac{V}{R}$, can be easily rearranged:

thus if we multiply both sides of the equation by R we get

$IR = V$, an alternative form of Ohm's law.

A third form exists: if we divide both sides of the last equation by I we get

$$R = \frac{V}{I}$$

Figure 3.2 may help to remind you of the three forms. They are all important.

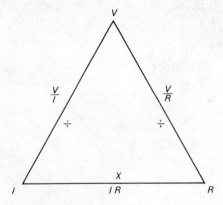

Fig. 3.2 To help remember Ohm's law

Examples 3.1 and 3.2
1. If a 50-V supply across a resistor causes 5 A to flow, what is the value of resistance?

Using $R = \dfrac{V}{I}$ $R = \dfrac{50}{5} = 10 \, \Omega$

2. If 2 A flows through a 6-Ω resistance, what is the p.d. across the ends of the resistor? (Note that the term used for a component having resistance is 'resistor'.)

Using $V = IR$ $V = 2 \times 6 = 12 \, V$

The standard symbol for a resistor is shown as a small rectangle and its resistance in ohms can be shown either inside the box or alternatively alongside. The older symbol for a resistor is a zig-zag line (Fig. 3.3). Both will be found in circuit diagrams.

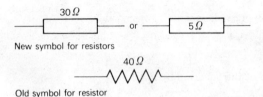

Fig. 3.3 Symbols for resistors

The symbol for an electric cell is a long vertical line alongside a shorter one. Figure 3.4 shows this. The long line is regarded as the positive plate and the shorter line as the negative plate.

Single cell   'Battery' of cells

Fig. 3.4

Batteries are a collection of cells connected together. Electric signal filament lamps or bulbs are usually shown by a circle with a cross inside (Fig. 3.5). The symbol for a switch is also shown.

Electric signal lamp   Switch

Fig. 3.5

Resistors in series

If a number of resistors are connected in cascade, i.e. in series, and a voltage is placed across the chain of resistors it is obvious that the same electric current flows through each (Fig. 3.6).

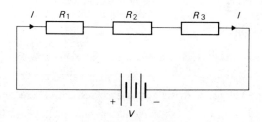

Fig. 3.6 Series connection

The p.d. across each resistor is given by Ohm's law. Thus the p.d. across R_1 is IR_1; across R_2 is IR_2; and across R_3 is IR_3.

The total p.d. across the three resistors is the sum of the individual voltages,

i.e. total p.d. $= IR_1 + IR_2 + IR_3$
$= I(R_1 + R_2 + R_3)$

This total p.d. is the voltage applied from the battery so that

$V = I(R_1 + R_2 + R_3)$

If the three resistors R_1, R_2 and R_3 are now replaced by an equivalent single resistor R_T having the same effective resistance then the same voltage must cause the same current to flow,

i.e. $V = IR_T$

Comparing these last two equations, the effective resistance in the first case is $R_1 + R_2 + R_3$ and in the second R_T.

Thus $R_T = R_1 + R_2 + R_3$

i.e. the total resistance of resistors in series is given by the sum of the individual resistances.

Resistors in parallel

Figure 3.7 shows three resistors R_A, R_B, R_C connected alongside each other. The arrangement is called a parallel connection.

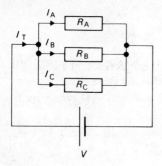

Fig. 3.7 Parallel connection

From the diagram the same voltage V exists across each resistor. Therefore the current in R_A is given by

$$I_A = \frac{V}{R_A}$$

In the same way $I_B = \dfrac{V}{R_B}$ and $I_C = \dfrac{V}{R_C}$.

The total current I_T flowing from the voltage source is the sum of the individual currents

i.e. $I_T = I_A + I_B + I_C$

Hence $I_T = \dfrac{V}{R_A} + \dfrac{V}{R_B} + \dfrac{V}{R_C}$

$= V\left(\dfrac{1}{R_A} + \dfrac{1}{R_B} + \dfrac{1}{R_C}\right)$

If an equivalent resistor R_T replaces the three parallel resistors then the same current must flow from the same applied voltage.

Thus $I_T = \dfrac{V}{R_T}$ or $V\left(\dfrac{1}{R_T}\right)$

It follows therefore, comparing the last two equations, that

$$\dfrac{1}{R_T} = \dfrac{1}{R_A} + \dfrac{1}{R_B} + \dfrac{1}{R_C}$$

i.e. the reciprocal of the total resistance of resistors in parallel is given by the sum of the reciprocal of the individual resistances.

Examples 3.3, 3.4, 3.5

1. Two 50-Ω resistors are connected: (a) in series; (b) in parallel. What is the resistance in each case?

    Series $R_T = 50 + 50 = 100\ \Omega$

    Parallel $\dfrac{1}{R_T} = \dfrac{1}{50} + \dfrac{1}{50} = \dfrac{1+1}{50} = \dfrac{2}{50} = \dfrac{1}{25}$

    Hence $R_T = 25\ \Omega$

    Note that resistors in parallel have a total resistance less than the resistance of any one resistance.

2. What value of resistance is required in series with a 10-Ω resistance to produce an effective resistance of 50 Ω?

    $R_T = 50\ \Omega \qquad R_T = x + 10 \quad$ where x = unknown resistance

    It follows that $x = 50 - 10 = 40\ \Omega$

3. What value of resistance is required in parallel with a 20-Ω resistance to produce an effective resistance of 12 Ω?

    $R_T = 12 \qquad \dfrac{1}{R_T} = \dfrac{1}{12}$

    $\dfrac{1}{R_A} = \dfrac{1}{20} \quad \dfrac{1}{R_B} = \dfrac{1}{x}$ where x = unknown resistance

    Thus $\dfrac{1}{12} = \dfrac{1}{20} + \dfrac{1}{x}\ $ from the equation $\ \dfrac{1}{R_T} = \dfrac{1}{R_A} + \dfrac{1}{R_B}\ $ for resistors in parallel.

    Rearranging $\dfrac{1}{12} - \dfrac{1}{20} = \dfrac{1}{x}$

    Putting the left-hand side on a common denominator

    $\dfrac{5-3}{60} = \dfrac{1}{x}$

    hence $\dfrac{2}{60} = \dfrac{1}{x}$

or $\frac{1}{x} = \frac{1}{30}$

It follows that

$x = 30 \, \Omega$

Series-parallel connections

Resistors can be connected in both series and parallel in the same circuit. Thus three 30-Ω resistors can be arranged in the various forms shown in Fig. 3.8.

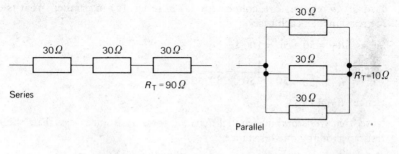

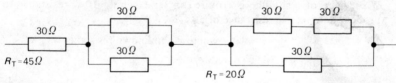

Fig. 3.8 Series-parallel connections

If they are in series

$R_T = 30 + 30 + 30 = 90 \, \Omega$

In parallel $\frac{1}{R_T} = \frac{1}{30} + \frac{1}{30} + \frac{1}{30} = \frac{3}{30} = \frac{1}{10}$

$R_T = 10 \, \Omega$

In series-parallel the two possibilities are

$R_T = 30 + 15 = 45 \, \Omega$ (why?)   or

$\frac{1}{R_T} = \frac{1}{30} + \frac{1}{60}$

$= \frac{2+1}{60} = \frac{3}{60} = \frac{1}{20}$

$R_T = 20 \, \Omega$

Bearing in mind that the final circuit need not have all the resistors in it, there are the following possibilities with three 30-Ω resistors:

10 Ω; 15 Ω; 20 Ω; 30 Ω; 45 Ω; 60 Ω; 90 Ω.

It is left to the reader to show how these values may be obtained.

Division of current

The current in any part of a circuit can be found by direct application of Ohm's law in one form or another. The following example will demonstrate how this can be done.

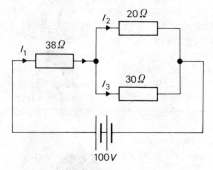

Fig. 3.9 (Example)

Example 3.6

Find the current in each part of the network shown in Fig. 3.9.

The resistance of the 20-Ω and 30-Ω resistors in parallel is found from

$$\frac{1}{R_p} = \frac{1}{20} + \frac{1}{30} = \frac{3+2}{60} = \frac{5}{60} = \frac{1}{12}$$

$\therefore R_p = 12$ Ω

The total resistance of the circuit is $R_{Total} = 38 + 12 = 50$ Ω

$$I_1 = \frac{V}{R_{Total}} = \frac{100}{50} = 2 \text{ A}$$

The p.d. across the 38-Ω resistor is given by $I \times R$

i.e. $2 \times 38 = 76$ V

This leaves $100 - 76 = 24$ V across the two parallel resistors

The current through each is given by $I = \dfrac{V}{R}$

Hence $I_2 = \dfrac{24}{20} = 1\cdot 2$ A

$I_3 = \dfrac{24}{30} = 0\cdot 8$ A

As a check I_1 should be the sum of I_2 and I_3

$I_2 + I_3 = 2$ A

Hence the answers check.

The resistance of different materials

Generally speaking electrical materials fall into three categories as far as resistance is concerned:

Conductors   e.g. copper, aluminium, silver
Insulators   e.g. glass, mica, quartz, paraffin wax, most plastics
Semiconductors   e.g. silicon, germanium.

The resistance of a sample of material, e.g. aluminium, depends upon the dimensions of the material. Several strands of aluminium in parallel will have a lower total resistance than a single strand. It follows that material of a large cross-sectional area has a lower resistance than similar material of a small cross-section.

    i.e. resistance decreases with increase in cross-sectional area.

Also several strands of similar wire in series will have a larger resistance than a single strand.

It follows that increase in length of material increases the resistance.

    i.e. resistance increases with length.

Mathematically we can say with resistance R

$R \propto \dfrac{l}{A}$ where l = length, A = cross-sectional area.

The sign $\propto$ means 'is proportional to'.

Resistance also varies with temperature — some materials (mainly in the conductor category) have an increase in resistance as the temperature rises. These materials are said to possess positive temperature coefficients of resistance.

Other materials (in particular semiconductors) show a decrease in resistance as the temperature rises. These materials are said to possess negative temperature coefficients of resistance.

Insulators, which conduct virtually no current at all, show often little change in resistance with temperature unless very high temperatures are encountered which change the physical structure of the material.

Linear and non-linear resistances

A linear resistance is one in which the current flow through it is directly

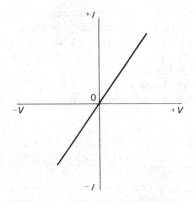

Fig. 3.10 I–V graph of a linear resistor

related to the voltage (p.d.) across it. A graph showing the variation of current with voltage is illustrated in Fig. 3.10. This shows a straight-line graph and moreover, when the polarity of the voltage is changed the relationship is still a linear one.

Due however to the temperature coefficient possessed by most materials many resistors do not have a linear relationship.

As current flows through a resistor, the resistor tends to get hot. As its temperature rises, its resistance changes. Figure 3.11 shows the current–voltage relationships for two electric lamps, the first with a tungsten filament, the second with a carbon filament. Tungsten has a positive temperature coefficient so that the current does not have the same increase at high voltages, where the current flowing is greater and hence the temperature is higher, as it does at low voltages.

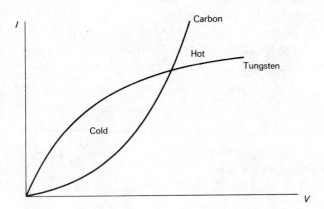

Fig. 3.11 I–V characteristics of two filament lamps

Carbon on the other hand has a negative temperature coefficient and the graph curves in the opposite direction. If we wish to maintain a constant resistance therefore we must: (a) choose material which has as small a temperature coefficient as possible; (b) keep the current through it as low as we can in order that its temperature does not rise unduly.

Filament lamps obviously make poor linear resistors. Some devices, while having non-linear characteristics, show distinct variations when the polarity of the voltage across them is reversed.

Figure 3.12 shows an example where with positive voltages the current increases rapidly with voltage but in a non-linear fashion. With the reverse polarity the current reaches a saturation value very quickly. This is typical of a rectifier which will be shown in Chapter 6.

One has to exercise care when dealing with non-linear resistors if Ohm's law is to be applied. Stricly speaking Ohm's law can be used only when dealing with linear resistors.

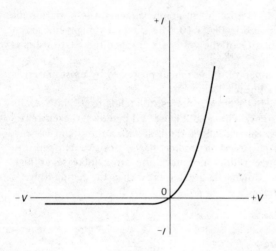

Fig. 3.12 Rectifier characteristic

Questions

Complete the following statements by indicating your choice 1, 2, 3 or 4.
(a) Potential difference is measured in
   1. amperes
   2. volts
   3. ohms
   4. length

(b) 1. $V = IR$
2. $V = \dfrac{R}{I}$
3. $R = \dfrac{I}{V}$
4. $I = \dfrac{R}{V}$

(c) Three 6-Ω resistors in series have a total resistance of
1. 2 Ω
2. 5 Ω
3. 6 Ω
4. 18 Ω

(d) Three 6-Ω resistors in parallel have a total resistance of
1. 2 Ω
2. 5 Ω
3. 6 Ω
4. 18 Ω

(e) Current through resistors in parallel is
1. always the same
2. proportional to the individual resistances
3. proportional to the reciprocal of the individual resistances
4. always different

(f) Silicon is
1. a conductor
2. a semiconductor
3. an insulator
4. a superconductor

(g) Semiconductors have
1. positive temperature coefficients
2. negative temperature coefficients
3. zero temperature coefficients
4. are affected only by a change in polarity of applied voltage

(h) Resistance of a material
1. increases with both length and cross-sectional area
2. increases with length but decreases with cross-sectional area
3. decreases with both length and cross-sectional area
4. decreases with length but increases with cross-sectional area

(i) A linear resistor can be described as
1. a resistance which is in a straight line
2. a resistance which behaves as a rectifier
3. a resistance which obeys Ohm's law
4. a resistance which fails to obey Ohm's law

Problems

1. Three resistors have resistances 20 Ω, 30 Ω and 40 Ω. List the various values which can be obtained from these three resistors when all or some are connected together: (a) in series; (b) in parallel.

   Solutions: (a) 20 Ω, 30 Ω, 40 Ω, 50 Ω, 60 Ω, 70 Ω, and 90 Ω.
   (b) 9·23 Ω, 12 Ω, 13·33 Ω, 17·14 Ω

2. Find the current through each resistor in the networks shown in Prob. Figs 3.1(a)–(c).

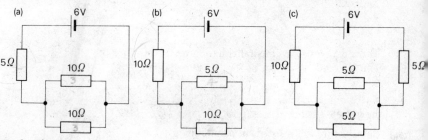

Prob. Fig. 3.1

   Solutions: (a) 0·6 A, 0·3 A, 0·3 A
   (b) 0·45 A, 0·3 A, 0·15 A
   (c) 0·342 A, 0·171 A, 0·171 A, 0·342 A

3. Find the value of R in each of the cases shown in Prob. Figs 3.2(a)–(e).

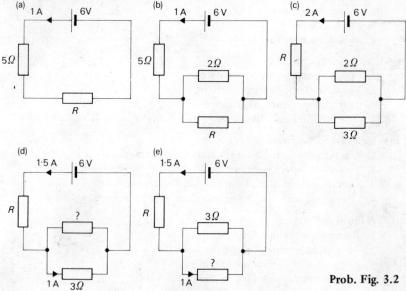

Prob. Fig. 3.2

   Solutions: (a) 1 Ω; (b) 2 Ω; (c) 1·8 Ω; (d) 2 Ω; (e) 3 Ω.

4. What is the e.m.f. E in the networks shown in Prob. Figs 3.3(a), (b)?

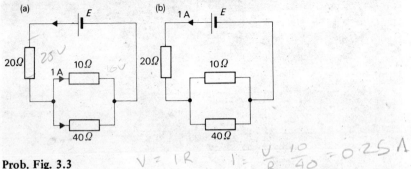

Prob. Fig. 3.3

Solutions: (a) 35 V; (b) 28 V.

5. If R is variable from 0 to 50 Ω, what is the range of current taken from the voltage source in each of the cases shown in Prob. Figs 3.4(a)–(c)?

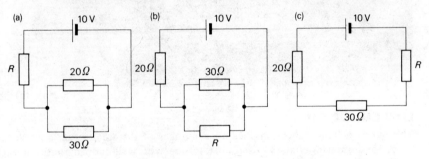

Prob. Fig. 3.4

Solutions: (a) 0·83 A, 0·16 A; (b) 0·5 A, 0·26 A; (c) 0·2 A, 0·1 A.

Chapter 4

Electronic systems

Introduction

The word 'system' can be applied to a number of different situations. We have, for instance, a railway transport system, with well-defined routes and stations. We each one of us have in our bodies several systems — a blood system for example involving the heart, the kidneys, arteries and veins, etc., a digestive system, and so on. In more scientific language we may regard systems as cognate or similar organisations or groups of objects which have boundaries sometimes clearly defined, occasionally ill-defined. Systems can have links, one with another, and are often interdependent. As an example the civilisation in the UK (a social and economic system) depends upon a transport system (rail, road, air and water transport), a communication system (post, telephone, video links, broadcasting), an educational system (primary, secondary, higher), a health service, etc.

Remove completely even a section of a system and very soon society as we know it begins to break down. A nationwide rail strike, for example, leading to a strike of transport drivers and the eventual withdrawal of the bus services means that many people would not be able to get to work; that fuel for the power stations would not arrive at its proper destination and we would be deprived of electricity; that food would no longer reach the shops; that foodstuff for the livestock on farms would be restricted; etc. Society would indeed be very different to the one we know now.

Systems can involve people, or be simply hardware, or can be a combination of both. Systems can be large or small. Figure 4.1 depicts a number of interconnected large systems.

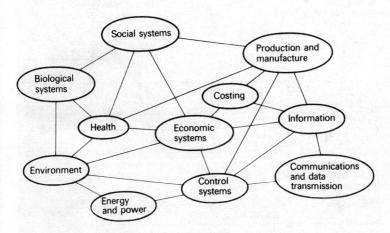

Fig. 4.1 Large systems which interact

Large systems can be broken down to smaller systems, or subsystems. A car is an example of an electro-mechanical hardware system having within it many subsystems. There is the engine — a largely mechanical system. There is the electrical system — ignition, alternator, head and side lights, etc. There is the transmission system — clutch, gear box and differential. There is the control system — steering wheel, foot pedals, gear lever — which must interact with the human driver.

A system involving people is illustrated in Fig. 4.2.

Systems have inputs and outputs and in the simplest case the system has a well-defined boundary and a single input and a single output (Fig. 4.3). Many of the smaller electronic systems fall into this category.

Electronic systems

Although a number of electronic systems are complicated they can invariably be broken down to smaller and therefore simpler systems with often a single input and a single output. Electronic systems can usually be better understood in their functioning when broken down in this way. It is not often necessary to know precisely what is inside each subsystem in order to know how it functions — that is, it is unnecessary to be aware of the individual components employed in each case. What must be known is the action that the system or subsystem produces on anything we put into it. In other words, we have to know the relationship between output and input. Such a way of regarding electronic systems is often referred to as a 'black box' approach.

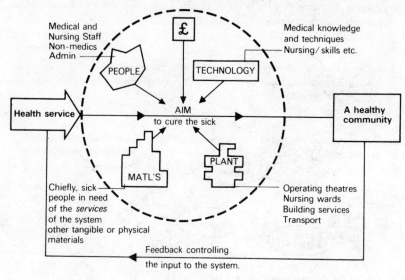

Fig. 4.2   A hospital system

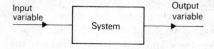

Fig. 4.3   A simple system

Amplifiers and attenuators

A simple electronic 'black box' system is shown in Fig. 4.4. The input voltage in this case is applied to a pair of input terminals (the input port) and the output voltage is available at the pair of output terminals (the output port).

If the input is a voltage e_i and the output is a voltage e_o the relationship between the two can be expressed mathematically as

$$e_o = A e_i$$

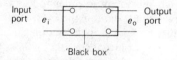

Fig. 4.4   A simple electronic system

'A' is the action produced by the black box and is sometimes called a transfer function.

If A is a constant greater than unity, e.g. the number 15, e_o is similar to e_i but 15 times as large and the black-box system is an amplifier. In this case, to be precise in the definition, it is a 'linear' amplifier, the word linear meaning that the output and input are identical in all respects except size.

If A is a constant less than unity, e.g. 0·1, the system is called an attenuator, e_o is now less than e_i; in fact e_o is $\frac{1}{10}e_i$.

Often in electronic systems one of the terminals of the input and one of the terminals of the output are each connected to an earthed point and are therefore common. We sometimes dispense completely with the earth connection in our black boxes when this happens (Fig. 4.5). This reproduces in effect Fig. 4.3.

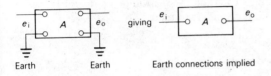

Fig. 4.5 Reduction to a single input and output

Other electronic black boxes

In many cases the transfer function A is not a constant and the output and input are no longer similar. This often happens with amplifiers where the input exceeds the designed value and we say that the amplifier is overloaded, and distortion occurs. Figure 4.6 shows such an example.

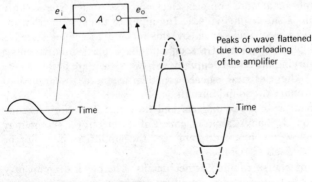

Fig. 4.6 Amplifier distortion

In certain electronic systems (especially logic circuits) the output is deliberately arranged not to be similar to the input. So far black boxes have been drawn as small rectangles. We can differentiate between black boxes which act in different ways by putting a label on each. Sometimes certain well-known black boxes are shown by accepted standard symbols. For example, a logic circuit known as a NAND (not–AND) gate is a square with the symbol & and a small circle (Fig. 4.7). A linear amplifier is often shown as

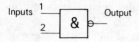

Fig. 4.7   A two-input 'NAND' gate

a triangle (Fig. 4.8). Where there is the slightest doubt of the identity or function of a black box it should be labelled. Normally the input of a black box is shown on the left and the output on the right in much the same way as one reads a line of print in a book from left to right.

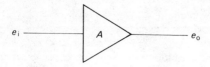

Fig. 4.8   An amplifier symbol

Block diagrams

Black boxes may be linked together to form a block diagram, i.e. the subsystems (black boxes) may be interconnected to form a more complicated system. An example of a more complex electronic system is a stereo 'hi-fi' record player. This is shown in Fig. 4.9. The input to the system from the record is changed to electrical signals by the pick-up and since stereo is involved twin amplifiers are needed throughout. Each black box denotes a 'stage' in the amplifying process. Eventually the two signals are passed to the two speakers and the electrical signals are then changed to mechanical vibrations which produce the sound outputs.

The total circuit is a block diagram and does not illustrate the components in each box. The physical arrangement of the electronic system may not necessarily follow the block diagram, i.e. the amplifiers may not be arranged in a line from left to right.

It is possible of course to add further details. The block diagram may include the power supply for the amplifiers (since they cannot operate directly from the supply mains) and the motor which spins the record (Fig. 4.10). There is a further input to the system other than the signal, and that is an electrical supply. This input should not feature in the output at all although in a badly designed amplifier may constitute an unwanted 'hum' in

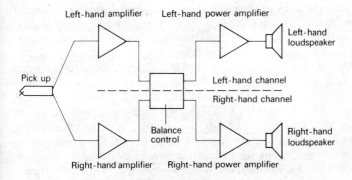

Fig. 4.9 A stereo amplifier — block diagram

the output. The 50-Hz* a.c. supply is a necessary part of the system without which it cannot function. The alternating voltage is transformed to a lower value and then changed to a direct voltage in the power supply unit. In practice a proportion of this electrical input is converted to the sound energy which comes out of the loudspeakers. The percentage which appears as required output is small, but this fact is not usually regarded as of great importance since only small amounts of electrical power are involved. In

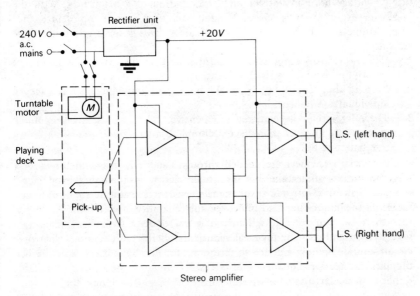

Fig. 4.10 Further additions to the system

---

*The electrical mains supply alternates 50 times per second. The supply frequency in 50 cycles per second or 50 hertz — abbreviated to 50 Hz.

other words, the efficiency of the system (wanted output power ÷ total input power) is low but is not considered a major factor. Where much higher powers are concerned efficiency becomes more important. With electronic systems, efficiency in the terms described is rarely of top priority.

The block diagram is essentially the first stage in understanding an electronic system. It does not generally indicate where the 'black boxes' can be found in the physical system nor does it necessarily imply that the circuit is broken down in the manner shown by the block diagram. Furthermore, it does not show how each individual box functions or what is contained in each box. However, should an electronic system fail to function correctly it is usually a relatively simple matter to locate which box or subsystem is not working as it should, and then either to repair or replace it. Modern electronic technology is moving rapidly to a position in which malfunctioning subsystems are located and replaced rather than repaired, and the design of the better electronic equipment is such that subunits can be easily replaced. In older equipment (especially those employing thermionic devices) the only part which is easy to replace is the valve itself and repair is more commonplace if components other than the valve are at fault.

While the full understanding of the internal details of a subsystem or black box is not usually necessary, some knowledge of the effect of the environment on its operation is required. Often high humidity and temperature have a deleterious effect. It is well to know too whether variations in the supply voltage have any effect on the operation of a black box. With the increasing use of the microcircuit the approach to the repair and maintenance of electronic systems has changed significantly. This is due primarily to three reasons.

1. The individual components in a microcircuit are extremely small and cannot be easily recognised or located.
2. Even if they could be, it would be almost impossible to replace an individual component.
3. The microcircuit is relatively inexpensive. It is thus usually very much quicker and cheaper to replace a complete microcircuit than to attempt to repair. See Fig. 4.11.

The design of microcircuits is obviously a specialist task which relatively few technicians undertake but the microcircuit is becoming increasingly familiar in many electronic systems. The considerable advancement of manufacturing techniques has also improved the reliability of electronic circuits so that there is often much less chance of a microcircuit (which might contain several thousand semiconductor components) giving fault than an electronic circuit containing only a dozen or so thermionic components. Frequently it is the interconnection between microcircuits which is the source of trouble rather than the circuits themselves.

Drawing block diagrams

While there are no laws laid down in the drawing of block diagrams there are

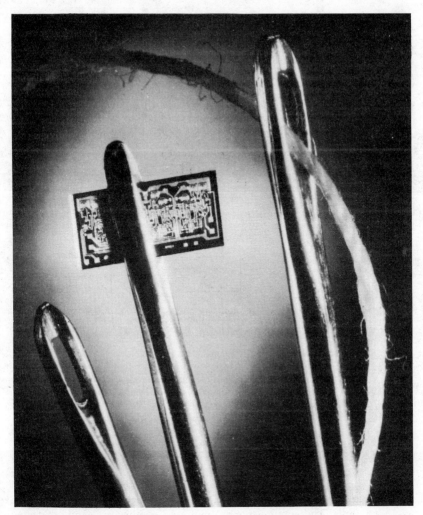

Fig. 4.11 A microcircuit

Photograph courtesy Mullards

certain generally accepted codes of practice. One had already been mentioned, namely, the input is usually drawn on the left-hand side and the output on the right-hand side of the diagram. A few more are:
1. Power supply lines, if included on a block diagram, are usually drawn with the unearthed line (whether positive or negative) in the top half of the diagram. The earthed line (if one exists) is usually drawn underneath the boxes.
2. Each box is rectangular and labelled except that sometimes amplifiers are shown as triangles and are then unlabelled. On occasions a symbol other

than a rectangular box or triangle is drawn to depict the contents, e.g. a loudspeaker.
3. Logic circuits are usually shown as square boxes but the symbol on the box describes the function of each unit.
4. Arrows are sometimes added to show the direction in which the signal moves through the system. This is sometimes necessary if signals are fed back from output to input (a situation known as feedback) or if a very complex system is considered.

A computer is shown in Fig. 4.12 in block diagram form.

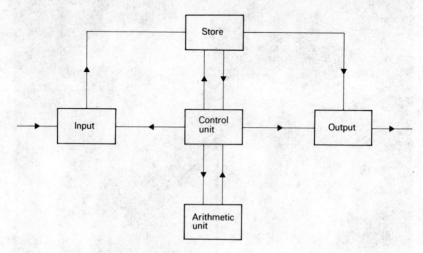

Fig. 4.12 Block diagram of a computer

Printed circuit boards

Today a number of electronic circuits involving discrete (i.e. independent and identifiable) components as opposed to the microcircuit are arranged on insulated cards. The components are mounted on the cards and connected together by a thin layer of copper etched on to the card. These cards are of two main forms:
1. In this case the card is designed for a specific purpose and the component tag ends are pushed through holes in the card and soldered in place. Providing the right components are in the correct position there is no need to check the circuit (except for poor soldered connections) as the interconnections between the components have all been arranged beforehand by the designer of the card. Often a card is a unit 'black box' of the block diagram. Figure 4.13 illustrates such an arrangement.
2. The card has narrow strips of copper etched in parallel on one side. The person using such a card has then to work out for himself how the connections to the various components are to be made. It is often

Fig. 4.13 Printed circuits

Photograph courtesy H.S.D. (Stevenage)

necessary to make breaks in the copper strips to isolate one component from another. See Fig. 4.14.

Both types of card are referred to as printed circuit boards. The first type is used in large-scale production. The second type is used in experimental situations where only one or two similar cards are in use. The tendency is to develop the first type from the second.

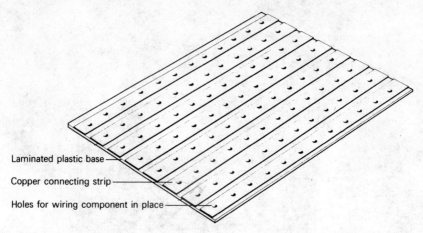

Fig. 4.14 Printed strip board

Summary

Summarising, there is a decreasing order of complexity in systems in general and electronic systems in particular.
1. The total system, e.g. a computer.
2. The subsystem, e.g. teletypes, store, memory, etc.
3. The individual 'black boxes' forming the subsystem, e.g. amplifiers, logic circuits.
4. The components within each 'black box', e.g. transistors, resistors, sockets, etc.

Usually 2 and 3 are illustrated by means of block diagrams.

When dealing with microcircuits very often the components are not revealed at all.

Questions

1. Produce a block diagram of a car showing clearly the various subsystems and the links between them.
2. A voltage wave of peak or maximum value 1·5 V is applied to two black boxes in turn. In each case the waveforms coming out are the same shape

as those going in but have peak values of 0·05 V and 22·5 V respectively. What do the black boxes represent?

3. The waveform applied to an amplifier is a rectangular wave of peak value 0·5 V. The amplifier has a variable gain but saturates at an output of 20 V. What is the peak value of the output at gain settings of 10, 20, 50, and 100? Draw also for each gain setting the input and output waveform for a triangular input wave of peak value 0·5 V.

4. The circuit shown in Prob. Fig. 4.1 is a two-stage amplifier, each stage having a gain of 15. Draw a block diagram to represent this circuit.

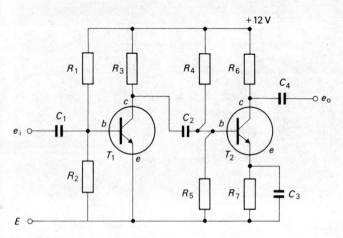

T_1 and T_2 are transistors (See chapter 6)
R_1-R_7 are resistors
C_1-C_4 are capacitors

Prob. Fig. 4.1

5. Show how the components in the circuit of the Prob. Fig. 4.1 can be connected together on a printed circuit board. Identify each component by its code number, e.g. R_2, C_1. Either type of printed circuit can be employed. An input and an output connection should be clearly shown as well as the d.c. supply leads.

Chapter 5

Signals and data transmission

Information

In engineering systems it is necessary to collect, store and transmit information of one sort or another. The information may be simply of the form that a particular condition, such as the required temperature of a furnace, has been attained. It might be in the form of some numerical value — for example, that so many components in a production process have been manufactured. It might be a much more complicated and lengthy item of information such as a report on the quality of a particular steel. Complex information is normally expressed in words, numbers and symbols in a written report and stored on paper in a filing system. It is usual to give it some form of address or reference number so that it can be easily found when needed. Large quantities of information are stored in books or pamphlets placed on library shelves. A method of address is used (such as the Dewey classification system) so that books can be easily located.

The problem nowadays is storing masses of information in a reasonably accessible form and in a reasonably economic manner. Frequently the information is reduced in physical size by using microfilm techniques. Occasionally the information is stored within a computer, again with a coded address so that it can be traced quickly when needed.

Often information is presented in a standard form as with a card index system so that a large amount of information can be seen at a glance especially by a trained observer. Sometimes the information is stored by punching holes in cards or tape or by magnetising metallised tapes or discs. The simplest form of information can be presented visibly by an indicator lamp which glows when a required condition has been reached, e.g. a green lamp might glow when the required temperature has been reached in a furnace. Numerical information can be presented by a counter display, e.g. the mileage indicator of a car. More complex information is displayed in words, numbers and symbols or by a diagram and photograph. Information stored in a computer on magnetic tape is normally fed into a teletype and displayed as typewritten characters.

Transmission of information

Information is often required to be transmitted some distance from the source. The method used to transmit the information depends upon how complex the information is, the distance over which it is to be transmitted and the speed at which the information is needed.

In the example just given the furnace might be located only a few metres from the control panel where the indicator lamp is situated. The connection from the temperature-sensing element at the furnace to the control panel is direct by means of conducting wires. There is no time delay in transmitting the information that the required temperature has been reached.

Information from a computer to a teletype might have to be transmitted several kilometres. Although each item of information is transmitted virtually instantaneously, it takes a significant time for a quantity of information to be presented by the teletype.

On occasions when there is no urgency in passing information from one place to another, it is possible to use the postal system and convey in a book or written report a considerable quantity of information with a delay of possibly several days (especially using second-class mail). Where it is necessary to have an exchange of information delays of several days may be unacceptable and one then has to use a telex system (or teleprinter) or the telephone.

Much depends upon the complexity of the information and the speed at which it is needed. Generally speaking, the greater the speed of transmission the more sophisticated is the method of transmission and the greater the cost involved, although the telephone in the UK is still a surprisingly cheap mode of communication in spite of recent increases.

Basic requirements of a transmission system

In communication systems the information required has to be obtained, coded, addressed, transmitted, received and then decoded by the recipient of the information.

The sequence of the postal mode of communication as an example is as follows:
(i) Information obtained.
A technical report on the testing of a sample of steel, for instance, is received from the materials and chemical laboratories.
(ii) Information coded.
The results of the test are expressed in words and symbols and put down on paper in a written report.
(iii) Information addressed.
The report is placed in an envelope addressed to the person requiring the information.
(iv) Information transmitted.
The report is posted and is now in the hands of the Post Office with a delay of 1 or 2 days (or more!).
(v) Information received.
The recipient receives the envelope and its contents.
(vi) Information decoded.
Recipient reads the reports and gleans the information (see Fig. 5.1).

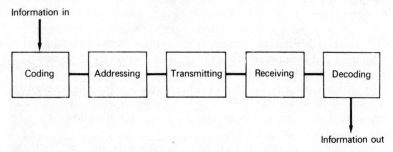

Fig. 5.1 Block diagram of an information transmission system

Other methods of transmission follow a similar pattern but the coding and decoding processes are often automatic and the transmission time is very short.

Types of signals

Broadly speaking signals fall into two categories — digital and analogue — and today the former is becoming of increasing importance. Digital signals are essentially of the ON/OFF type and are sudden changes in level of voltage (or current). As only two levels are involved the signal is referred to as a binary signal.

Analogue signals on the other hand are variations in voltage (or current) with time which follow rather more complicated patterns and have a whole range of levels.

The Morse code is one form of digital signal and the variations of voltage along a pair of telephone lines caused by the transmission of a conversation between two persons is an example of an analogue signal (Fig. 5.2).

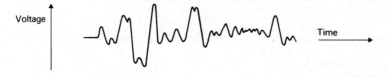

Fig. 5.2 An analogue signal

Digital signals require an associated code in order that intelligence may be transmitted. The dots and dashes of the Morse code can be translated first into letters and then words (Fig. 5.3).

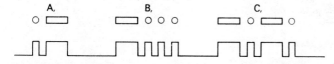

Fig. 5.3 The Morse code

Analogue signals do not normally require a code unless the language being employed (English, French, German, etc.) is regarded as a code.

Digital signals employ rectangular pulses which can be variable in length (as with the Morse code), or can use square waves of uniform duration which are joined in a coded sequence corresponding to the message being transmitted. If the latter type is used then the presence of a voltage is referred to as 1 (ONE) and the absence of a voltage as a 0 (ZERO). The 1's and 0's are

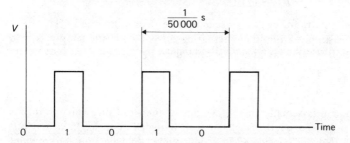

Fig. 5.4 The series of binary digits, bit-rate 100 kilobits/second

called binary digits or more briefly **bits**. The bit-rate is the number of ones or zeros per second and this can be very high. If ones and zeros are transmitted alternately the waveform is that shown in Fig. 5.4. (*Note:* This is a return-to-zero binary signal. There are other forms in use.) For the case where the number of bits per second is 100 000 the bit-rate is 100 kilobits per second.

The analogue signal

The simplest form of analogue signal is the sinewave (Fig. 5.5). The variation of voltage (or current) with time varies in a prescribed mathematical expression given by the equation

$$v = V_m \sin \omega t$$

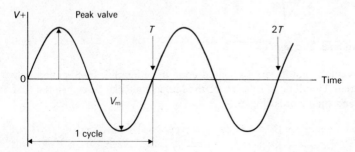

Fig. 5.5 The sinewave

where v is the voltage at any instant
V_m is the peak value of the sinewave
ω (the lower case Greek letter *omega*) is related to the number of complete waves or cycles occuring in one second
t is the time in seconds.

The wave form repeats itself after an interval T called the periodic time. The number of complete cycles per second is the frequency f.
f is related to the periodic time T by the relationship

$$f \text{ (cycles per second)} = \frac{1}{T \text{ (seconds)}}$$

The present-day term for cycles per second is **hertz** or more briefly Hz.

Thus the supply from the alternating-current domestic supply is at a frequency of 50 Hz and has a periodic time given by

$$T = \frac{1}{f} \quad \text{i.e. } T = \frac{1}{50} \text{ or } 0.02 \text{ second.}$$

The term **sin** is actually an abbreviation of the word **sine** which in turn is a property possessed by the angle in a right-angled triangle. If an angle θ (Greek letter *theta*) is considered in a right-angled triangle (Fig. 5.6) of sides h, p and b, then $\sin \theta = \frac{p}{h}$

This ratio is called a trigonometrical ratio and when θ is small, $\frac{p}{h}$ is small.

When θ is approaching 90°, $\frac{p}{h}$ approaches the value 1. Thus $\sin \theta$ varies from 0 to 1 as the angle θ varies from 0° to 90°. Tables are available giving values of

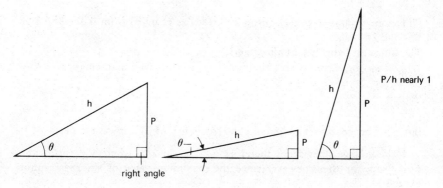

Fig. 5.6 $\sin \theta = \dfrac{p}{h}$

sines from 0° to 90°. It is possible to have sines of angles greater than 90° but the maximum value can never exceed unity. Sin θ decreases from 1 to 0 as θ varies from 90° to 180°. Between angles of 180° and 360° the value of sin θ varies from 0 through a maximum of -1 when θ is 270°, and back to 0 when θ is 360°. The cycle is repeated when θ exceeds 360°.

If the value of sin θ is plotted to a base of angle θ a graph is produced which is the sinewave already shown.

Angles can be measured in degrees or radians. A radian angle is obtained by drawing an arc of any convenient radius r (Fig. 5.7) and the angle in radians is then the arc length $ab \div$ radius r.

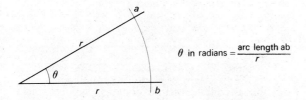

θ in radians $= \dfrac{\text{arc length } ab}{r}$

Fig. 5.7 Radian measurement

It follows that the arc length of an angle of 360° is $2\pi r$, the circumference of a circle of radius r.

Thus 360° (in degrees) $= \dfrac{2\pi r}{r}$ (in radians)

Thus 360° $= 2\pi$ radians

$\quad\quad$ 90° $= \dfrac{\pi}{2}$ radians

1 radian $= 57 \cdot 7°$

Returning to the sinewave of voltage $v = V_m \sin \omega t$, the angle θ must correspond to ωt.

When one complete cycle is produced, ωt has changed from $0°$ to $360°$, or through 2π radians.

The time t is then the periodic time T.

Hence $\omega T = 2\pi$

or $\quad \omega = \dfrac{2\pi}{T}$

but $\dfrac{1}{T} = f$ hence $\omega = 2\pi f$ where f is the frequency of the sinwave in hertz.

The voltage sinewave is therefore $v = V_m \sin 2\pi ft$ where V_m is now the peak value of the sinewave (since the maximum value of $\sin 2\pi ft$ cannot exceed 1).

The domestic mains electricity supply in the UK has a peak value of about 340 V and has a frequency of 50 Hz.

Its mathematical equation is therefore

$v = 340 \sin (2\pi \times 50t)$
$ = 340 \sin 314t$

The value of v can vary from 0 V to +340 V or 0 V to −340 V throughout the cycle. This is obviously different to the digital signal which has only two values.

More complicated analogue signals are built up from the addition of sinewaves of varying amplitudes and varying frequencies. Speech is typically composed of sinewaves of frequencies in the range of about 40 Hz to about 4 kHz (4000 Hz). Good quality 'hi-fi' involves frequencies of five or six times this figure and television signals need frequencies of up to 5 MHz, i.e. 5 million hertz.

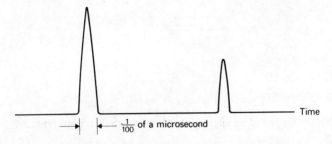

Fig. 5.8 Pulses in a radar system

A third type of signal is employed in certain cases, radar being a typical example. Here the signal is neither a repeated square nor a sinewave or mixture of sinewaves, but a short-duration pulse (Fig. 5.8). If the pulse is repeated at regular intervals the term pulse repetition frequency (p.r.f.) or pulse rate is used. In pulse systems the signal is at zero level for much of its time. The important part of a pulse system is when the pulse occurs and not the amplitude of the pulse.

Bandwidth

The range of frequencies occupied by a signal is known as the bandwidth and it is advantageous to produce signals which have as small a bandwidth as possible. Generally speaking the bandwidth of a transmission system is limited, and it is necessary therefore to ensure that the signal being transmitted is within this limited bandwidth, otherwise some of the signal may be lost.

Analogue signals may occupy a significant bandwidth. Digital signals may on the other hand occupy a relatively small bandwidth. The reason for this is that in the digital case only one of two alternatives exists — the current is ON (corresponding to 1) or OFF (corresponding to 0).

The signalling rate can be increased in the digital system by simply increasing the bit-rate (i.e. the number of square waves per second). In the analogue signal it is slightly more difficult. It would be possible to record a message which occupied a bandwidth of from 40 Hz to 3·6 kHz (say 4 kHz) on a tape recorder running at 5 cm/s and then to transmit the message at a tape speed of 20 cm/s. Although this transmits the message four times as quickly the bandwidth requirements have also increased by a factor of four and are now 16 kHz.

It is possible to transmit a picture by scanning the picture in a series of lines in much the same way as reading the lines of print in a book. The B.B.C. for example uses a 625-line scan in its T.V. picture. Different light tones in a monochrome picture (black and white) are transformed into voltages perhaps 0·5 V corresponding to 'white' and 0·1 V corresponding to 'black'. This is an analogue signal. The varying voltages corresponding to the light tones are then transmitted in sequence. If the picture is to be transmitted in a few milliseconds then a high bandwidth is needed (5 MHz). If the picture can be transmitted over several minutes then it may be necessary to provide a bandwidth of only 5 kHz (Fig. 5.9). The transmission of a single picture by this process is called facsimile transmission.

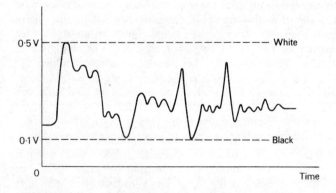

Fig. 5.9 T.V. monochrome signal (one line)

The bandwidth required for a signal depends upon the complexity of the information contained in it and the speed at which it is to be transmitted. Increase either and the bandwidth needed goes up.

Modulation

It is sometimes necessary to move a signal bodily from one band of frequencies to another. For example, as speech frequencies range from about 40 Hz to 4 kHz they occupy a bandwidth of nearly 4 kHz*. It is possible to shift the frequency spectrum of 0·04 to 4 kHz to a new spectrum still occupying the same bandwidth, such that the new signal might extend from, say, 100·04 kHz to 104 kHz. By this method it may be possible to transmit several signals simultaneously along a pair of lines. Thus we could have in a system having a total bandwidth of 100 kHz several speech channels, each requiring 4 kHz bandwidth but separated on a frequency base from an adjacent channel by, say, 6 kHz. By this process it is possible to have ten speech channels simultaneously transmitted along the same cable. This is known as frequency division multiplexing (Fig. 5.10).

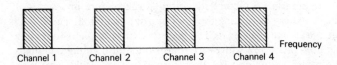

Fig. 5.10 Frequency division multiplexing

In order to achieve this frequency shift it is necessary to 'modulate' the signal. There are a number of alternative modulation methods. One is called amplitude modulation.

A relatively high frequency (say 80 kHz) is used as the signal 'carrier' and the speech signal of up to 4 kHz is placed on this carrier, resulting in the amplitude of the carrier varying with the applied signal (Fig. 5.11). It can be shown mathematically that this results in frequencies equal to the carrier frequency ± signal frequency. Thus in the example given, an 80-kHz carrier and a signal frequency of 4 kHz would result in frequencies of 76 to 84 kHz in addition to the carrier at 80 kHz. This has unfortunately doubled the original bandwidth and illustrates why it is necessary to have relatively large gaps between the various carrier frequencies in a frequency division multiplex system.

In radio transmission a similar process is used and the carrier frequency is the frequency to which the receiver has to be tuned. The receiver circuit has a frequency response which then embraces the signal frequencies above and below that of the carrier.

The frequencies in the band occupied by the carrier frequency f_c and the

---

* The Post Office allocates a bandwidth of 3·4 kHz to each of its telephone channels.

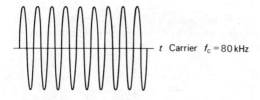
t Carrier $f_c = 80\,\text{kHz}$

t Signal $f_c = 4\,\text{kHz}$

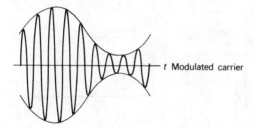
t Modulated carrier

Fig. 5.11 Amplitude modulation — waveforms

signals above the carrier frequency ($f_c + f_s$ (signal)) are located in the upper sideband while the other signals below f_c are in the lower sideband (Fig. 5.12). It is possible to use only one sideband for transmitting signals but this does result in a significantly more complex system.

Filtering and demodulation

With the frequency division multiplex system described it is necessary to be able to separate the different signals occupying different parts of the frequency spectrum. It is therefore necessary to have filter circuits which accept a band of frequencies centred around the various carrier frequencies.

If the first carrier has a frequency of 10 kHz then the first filter must separate out 10 ± 4 kHz, i.e. it must deal with a range of frequencies from 6 kHz to 14 kHz. The next channel occupies the band 20 ± 4 kHz, i.e. 16 kHz to 24 kHz, and so on.

Each filter must have the property of passing a band of frequencies and rejecting all others. Such an arrangement is called a 'band pass' filter. The required bandwidth is 8 kHz in each case and the centre frequency is located at 10 kHz, 20 kHz, 30 kHz, etc. (Fig. 5.13).

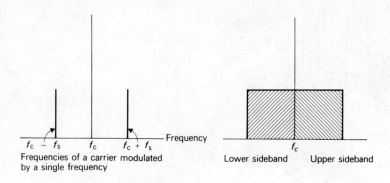

Fig. 5.12   Amplitude modulation — sidebands

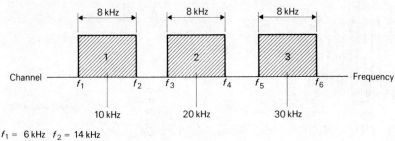

$f_1 = 6$ kHz $f_2 = 14$ kHz
$f_3 = 16$ kHz $f_4 = 24$ kHz
$f_5 = 26$ kHz $f_6 = 34$ kHz

Fig. 5.13   A frequency division multiplex system based on carrier frequencies of 10 kHz, 20 kHz, 30 kHz

When the various channels have been separated out it is still necessary to recover the original signal. The original signal occupied a band of frequencies from 0·04 kHz to 4 kHz and therefore it is necessary to remove the carrier from the received signal. This is fortunately achieved relatively simply by means of a rectifier (Fig. 5.14). The rectifier (a diode) allows passage of current in one direction only and thus removes the negative half of the modulated wave. The positive half has an average value which is somewhere around half-way between the peak of each individual wave of the received carrier frequency and zero. The audio system including the loudspeaker or headphones which follows the rectification process can only respond to the average value (i.e. the required signal). The system can be improved further by the use of a small capacitance across the resistor R_L. This charges up to the peak value of each carrier frequency pulse and thereby increases the size of the audio signal.

The process of removing the carrier is called demodulation.

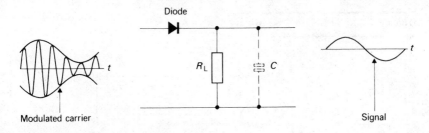

Fig. 5.14 Demodulation process

An identical process has to be employed in a radio receiver. The receiver is tuned to a particular frequency corresponding to the particular transmission which we wish to receive. Only frequencies in the required spectrum are then amplified. The carrier is removed by a similar demodulation process to the one described and the wanted signal is then after further amplification passed to the loudspeaker. The block diagram Fig. 5.15 shows this.

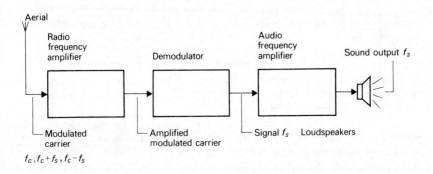

Fig. 5.15 Simple radio receiver

The digital signal

As mentioned earlier, the digital signal is becoming of increasing importance in modern transmission systems. It is as well to delve more deeply into the subject. The Morse code has already been referred to as one example of a digital or binary signal. Although it is still employed many binary signals today consist of a series of pulses of uniform length representing 1's (ONES) or 0's (ZEROS). It is the sequence of 1's and 0's which conveys the message. Take for example the decimal or denary numbers 0 to 9. These can be converted to binary numbers having only 0's and 1's. Thus

Denary		Binary
0	=	0
1	=	1
2	=	10
3	=	11
4	=	100
5	=	101
6	=	110
7	=	111
8	=	1000
9	=	1001

Denary numbers are in powers of 10 (units, tens, hundreds, thousands, etc.). Binary numbers are in powers of 2.
It is possible to change denary numbers into binary by a continuation of the above process.

Denary	Binary							
	64	32	16	8	4	2	1	(powers of 2)
Thus 83 =	1	0	1	0	0	1	1	
i.e. 64 + 16 + 2 + 1								

One thing should be obvious — numbers in binary use more digits than numbers in decimal or denary. One way to avoid too lengthy a binary number is to employ a slightly different code.

A denary number can be expressed in units, tens, hundreds, etc. Each may then be converted to binary, and a maximum of 4 binary digits are needed for each denary digit.

Take for example the number 7308
         7 thousands
         3 hundreds
         0 tens
         8 units
In binary 7 = 111
         3 = 11
         0 = 0
         8 = 1000

Often we 'round' these binary numbers into groups of equal length by adding zeros.
Thus   7 = 0111   (the first zero is redundant)
         3 = 0011
         0 = 0000
         8 = 1000   (no additional zero is needed here as 4 'bits' are already present)

Thus 7308 becomes 0111, 0011, 0000, 1000.
Such a coded system is known as binary coded decimal or B.C.D.
There are obviously many alternative codes which can translate denary numbers into binary.

Letters of the alphabet can also be expressed in a binary code. We could start at the binary number 10000 to represent A in order to avoid confusion with the numbers 0–9.
Thus one method of representing the alphabet would be

A = 10000
B = 10001
C = 10010
D = 10011
E = 10100
etc.

In binary language we use the term **word** to denote the number of bits constituting a number or a letter. Thus in the systems described above the word length for a denary number is four bits and for a letter of the alphabet five bits.

Such a coding has obvious limitations. If the sequence of binary coding is carried through it will be quickly found that only fifteen letters can be spelt, since 11111 is the highest number in binary which uses only five 'bits'. This corresponds to the denary number 31, but binary numbers up to 1111 have already been used for denary numbers. A rather more complicated code is necessary for the combination of ten denary numbers and the twenty-six letters of the alphabet. Codes which deal with both numbers and letters are called alphanumeric codes.

Comparisons of different transmission systems

A transmission system is composed of a number of essential components in addition to the sender and receiver of the messages. There is first the necessity of preparing the message to be sent. In a written communication the information is put down in words on paper. (This is essentially a process of coding — letters are arranged into words and words into sentences. If we cannot write, i.e. we do not know the code, we cannot easily commit the information to paper in a way that will be easily understood.) The message must now be converted to a form that can be conveyed by the transmission system. In a telephone system the differences in air pressure produced by the words we speak into the mouthpiece of the telephone cause electrical signals to be produced, that is, changes in current in the transmitting set. The message must now be addressed to the recipient and transmitted.

In the telephone system we dial the number we require. When sending a letter we place the message in an addressed envelope. In a radio system we may beam the radio signals by means of a specially designed aerial (or antenna) to the receiving station.

A carrier must be supplied to convey the message. In the case of the letter the Post Office provides or hires the physical means of transporting the letter (van, rail, plane). In the telephone circuit the wires 'carry' the information. In the radio system a high-frequency sinewave (the carrier) is used which is modulated by the signal. The transmission time is extremely fast in the case of radio (almost instantaneous). There is a delay of a few milliseconds with the telephone. A letter may take two or more days. At the receiving end the carrier is removed — demodulated in the radio system. The message must now be converted into a form which can be detected by the receiver (electrical signals changed to variations in air pressure in the headphones) and decoded if necessary. The message is then read or heard and hopefully understood by the recipient. Figure 5.16 compares four systems of information transmission. Only the last example (d) uses the terms 'coding', 'modulation' and 'demodulation' in their normally accepted sense by the various stages in any such system can be ascribed these titles.

In each case there is a time delay between sending and receiving the information. There is little to choose between (b) and (d) since the section of the system producing the greatest delay is the transmission section. Radio waves travel at 300 million metres per second, whereas telephone messages travel at about $\frac{1}{1000}$ of this speed. The sender and the receiver have to be very far apart before the time lag is significant in either case.

System (c) has the same transmission delay as (b), i.e. extremely small, but it takes some time to code and decode the message since the keys on the teleprinter have to be pressed in sequence and this depends upon the skill of the sending-end operator. At the receiving end, the digital signals have to operate a similar teleprinter and since the action is again sequential (and moreover mechanical) it takes a little while for the message to be typed out.

For two-way conversation (b) and (d) are obviously to be preferred.

Although the first method (a) is the slowest, a great deal of information can be conveyed in one transmission, e.g. a book can be sent by post. Information can also be conveyed in a diagram or photograph which cannot be readily conveyed by the other systems (b), (c) and (d) unless the picture or diagram is itself divided up into small sections (like the lines of print in a book) and transmitted sequentially. This process (facsimile) takes a little time in both coding and decoding.

Transmission errors

It is possible in the case of any transmission to produce errors. The written message in case (a) may not be clear or may contain ambiguities. The spoken message in (b) and (d) may equally well be misunderstood or words may not be heard. In (c) the operator may press the wrong keys, or malfunctioning of the receiving teleprinter may cause the wrong letters to be printed. Fortunately an odd mistake here and there causes the recipient little difficulty in understanding the sense of a message and transmission errors can normally be easily corrected.

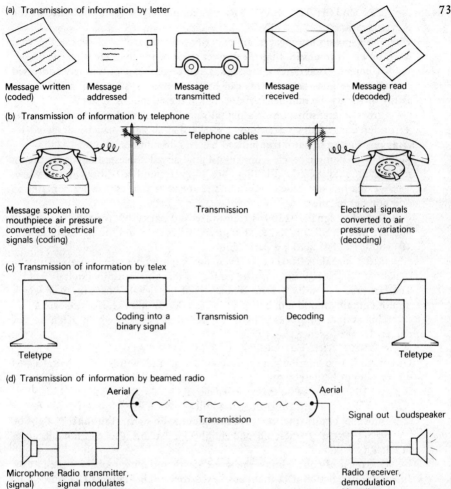

Fig. 5.16 Different information transmission systems
    (a) Transmission of information by letter
    (b) Transmission of information by telephone
    (c) Transmission of information by telex
    (d) Transmission of information by beamed radio

For example, the telex message
    CANNOT CAKCH THA LASK TRAEN
can be readily understood as CANNOT CATCH THE LAST TRAIN although four errors have occured. However, the omission of three letters will change the whole sense of a message, e.g.

CAN CATCH THE LAST TRAIN completely negates the sense of the previous message.

When this possibility occurs an experienced operator will repeat the vital word or words to ensure that the correct sense of the message is received.

Of course, sometimes a message is completely lost or wrongly addressed and therefore goes astray. This can happen in any system, e.g. a letter goes astray or the wrong number is connected in the telephone system.

Errors in transmission of a digital signal can often be detected by using 'error-detecting codes'. Spurious 1's in a digital signal can be detected by arranging that each word transmitted has an odd number of ones.

If, for example, each word in a digital signal consists of six bits then a seventh bit may be added if there are an even number of ones in the word. These extra bits are called 'parity bits' and ensure that each word contains an odd number of ones.

Thus if a signal 101100 is transmitted no parity bit is needed as an odd number of ones already exist. The 'parity bit' is zero and the signal sent out is 1011000. The last zero is a parity bit.

If the signal is 100111 an even number of ones are contained in the signal. A 'one' parity bit must be added so that the transmission is 1001111. The last one is a parity bit and is not part of the signal. Words received which do not contain an odd number of ones are in error and therefore rejected.

This system produces gaps in the received message which often do not detract from the sense conveyed.

e.g. HA-E ARR-V-D AT STA-ION A-D AM AWAIT-NG TAXI

Of course if two spurious ones are included in the 'words' of a digital signal errors will not be detected,

e.g. 1000110 may be the required signal
(including the parity bit)

If by mistake two further ones are introduced the signal transmitted might be 1101110. As 5 ones (an odd number) exist this also will not be recognised as a mistake.

One must of course know the word length and must recognize the last bit as an 'odd parity bit' which enables errors to be detected. Figure 5.17 illustrates some of the points made.

It is possible to produce codes which not only detect errors but also correct errors. This is however outside the scope of this book.

Summary

Information can be provided in a number of different forms but in transmission there are two basic types — analogue and digital. The simplest form of analogue signal is the sinewave. Analogue signals consist of a number of sinewaves of different frequencies. It is possible by a process of amplitude modulation to shift the signal from its original frequency range to a new one.

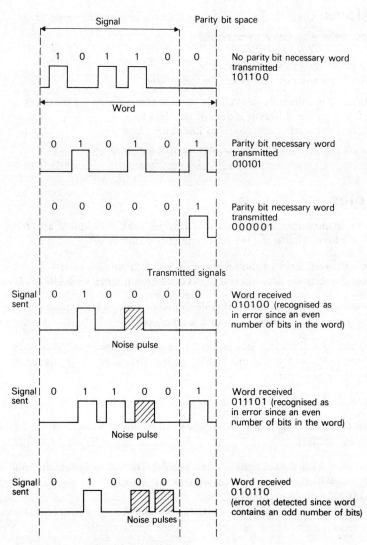

Fig. 5.17 The error-detecting code

This is called frequency division multiplexing. The digital signal usually occupies a smaller bandwidth than an analogue signal. Digital codes use the binary sequence and it is possible to produce both letters as well as numbers in a binary code. Errors can occur in transmission but, by using 'parity bits', it is possible to detect errors.

Questions

1. Compare the time taken to transmit
   (a) a twenty-word message
   (b) a two-thousand word message by:
   (i) letter and post; (ii) telex; (iii) telephone.
2. Compare approximately the relative costs of transmitting the messages in Question 1 by the different modes of transmission:
   (a) from London to York; (b) from London to New York.
3. What are the carrier frequencies of the three main T.V. channels?

Examples

1. Express in mathematical form a sinewave of peak value 50 V and frequency 1 kHz. What is the periodic time of this sinewave?

2. What is the frequency of the sinewave $i = 5 \sin 6280t$?
   What is the shortest time interval between the point when $i = 0$ and $i = 5$?

3. With the aid of trigonometrical tables, draw to scale two complete waves of the sinewave $v = 20 \sin 314t$ using a time axis.

4. A carrier $f = 250$ kHz is amplitude modulated by two sinewaves $v = 2 \sin 3140t$ and $v = 1 \sin 31400t$. What frequencies are produced in the modulated carrier?

5. Express the following denary numbers in binary: 15; 23; 64; 79.

6. Express the following in B.C.D. using a word length of 4 bits:
   4328; 1491; 2001.

7. A system uses a digital signal of word length six bits including the odd parity bit. Which of the following words are in error?
   (a) 110011   (b) 100011   (c) 000001
   (d) 111111   (e) 011010   (f) 101010

8. Using the code A = 1, B = 2, C = 3, etc., express the letters of the alphabet from A to O in binary form.
   Hence decode the following and answer.
   1111, 1110, 0101,
   0001, 1110, 0100,
   1110, 1001, 1110, 0101
   1101, 0001, 1011, 0101

9. A code uses an 'odd parity bit' as the final digit in a 5-bit word, but otherwise employs the same code as in Question 8. Decode the message.
00010
01100, 01011, 00111, 10000, 11100, 10011, 00111,
       10011, 00010 11100,
11100, 01011, 01011, 01000, 01111,
11110, 11111,
10110, 11100, 11111, 10001,
10001, 11111, 11010, 01011, 10001, 10000, 10011,
       11100, 01110,
00010, 00100, 11111, 11000, 11110,
00111, 11111, 01000, 01011, 10111.

Chapter 6

Power supplies and amplifiers

Introduction

Electronic circuits require electrical power in order to operate. This can be derived from batteries, from the domestic electricity supply mains or from a suitably designed generator driven by hand, by the wind or from some form of internal combustion engine. These latter power sources are examples of energy transformation — from natural sources (human, climatic or fossil fuel) to electrical energy.

In general the electrical energy must be direct and not alternating (i.e. d.c. and not a.c.) so that if the a.c. domestic mains supply is used it must first be converted from alternating to direct current. The process of changing from alternating current to direct current is called rectification and either thermionic or semiconductor devices may be used in the process.

The thermionic diode

The simplest form of thermionic device is the diode valve, so called because it has two electrodes. It uses basically the 'Edison effect' and has been mentioned already on page 3. The construction of a modern type of thermionic diode is illustrated in Fig. 6.1. It consists of a thin filament of wire (the

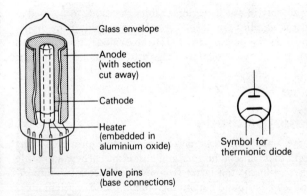

Fig. 6.1 The construction of a modern thermionic diode

heater) which is surrounded by, but insulated from, a thin nickel tube (the cathode) which is coated with a material containing barium and strontium oxides. When operating normally a voltage usually in the range of 4 V to 6 V is applied to the heater causing a current of perhaps 0·1 A to 0·3 A to flow. This raises the temperature of the cathode to around 1000 °C causing a copious emission of electrons from its surface. The heater supply can be alternating or direct current.

Surrounding the cathode assembly is the anode. This is another cylinder of nickel of greater diameter than the cathode. All this is enclosed in a glass envelope from which the air is excluded. Connections to the heater, cathode and anode are taken out through the base via pins.

When the cathode is heated and the anode is maintained at a positive potential to the cathode, emitted electrons cross the gap between cathode and anode, and current flows (in the opposite direction). When the anode is negative to the cathode, current ceases as the negatively charged electrons are repelled by the negatively charged anode.

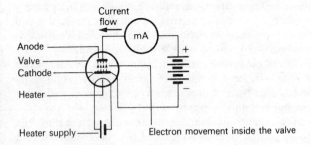

Fig. 6.2 Action of the thermionic diode

The action of the diode is illustrated in Fig. 6.2. The device described is called an indirectly heated thermionic diode. The fact that current between anode and cathode flows only in one direction enables the diode to be used as a rectifier. Two electrical supplies are needed for the operation of the diode: (i) a low voltage to provide the heating power for the cathode (this is termed a low-tension supply); (ii) a much higher voltage for the anode (called the high-tension supply).

The semiconductor diode

Semiconductors, as the name implies, have a conducting property lying somewhere in between that of conductors and insulators. They are made usually of germanium or silicon, the latter becoming of increasing importance. Older equipment may be found employing copper oxide or selenium as semiconducting material.

Modern semiconducting diodes employ silicon or germanium which is 'doped' by arsenic, antimony, boron or indium. At the risk of oversimplification, the way a semiconductor works is as follows.

The semiconductor material is first of all made extremely pure and allowed to form in a crystalline state. Under these conditions the material conducts electricity only slightly. Unlike conductors, which conduct by means of electron movement, semiconductors in the pure state conduct by means of electrons (negative charges) and 'holes' (positive charges). The name 'hole' is derived from the fact that a positive charge is created by a vacancy or gap in the crystal structure.

These electrons and holes both contribute to the flow of electric current. They are called 'minority-charge carriers' and increase in number with the temperature of the material. At a given temperature similar samples of pure silicon and germanium show dissimilar abilities to conduct electricity, silicon having the lower conducting quality.

A very small amount of impurity is now added to the pure semiconductor — arsenic or indium in the case of germanium.

Arsenic impurity causes a relatively large number of additional free electrons to be produced. These are then referred to as 'majority-charge carriers' and largely swamp any minority carriers present. The material is said to be 'doped'. The doped or impurity semiconductor now has a surfeit of mobile negative charges and is referred to as an 'n-type' semiconductor.

If indium instead of arsenic is added to germanium then a similar process ensues but this time relatively large numbers of free holes are produced as 'majority-charge carriers'. This type of doped semiconductor now has excess mobile positive charges and is termed a 'p-type' semiconductor.

To summarise, an n-type semiconductor has mainly negative mobile charges present but does contain a small number of positive mobile charges (minority carriers). A p-type semiconductor has mainly positive mobile charges present but does contain a small number of negative mobile charges (minority carriers). See Fig. 6.3.

When a pn junction is formed, a certain migration of the mobile carriers takes place — positive charges diffusing to the n region and negative charges

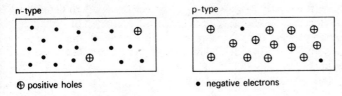

Fig. 6.3 n- and p-type semiconductors

to the p region. This causes a small potential barrier to be produced across the junction which prevents further movement of charges (Fig. 6.4). The application of an external voltage of the same polarity as this barrier merely

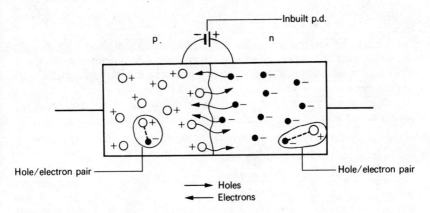

Fig. 6.4 The pn junction showing majority- and minority-charge carriers

enhances the effect and the only current which can flow across the junction is due to the minority carriers and is consequently very small (Fig. 6.5). The pn junction is said to be reverse biased under these conditions. If the polarity of

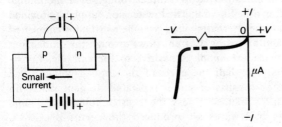

Fig. 6.5 Reverse biased pn junction

the external voltage is reversed the potential barrier is removed and the majority-charge carriers can readily cross the junction. In consequence a relatively large current flows (Fig. 6.6). Under these conditions the junction is said to be forward biased. The pn junction obviously behaves as a rectifier,

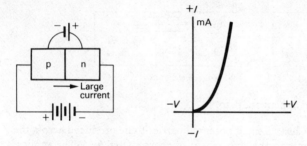

Fig. 6.6 Forward biased pn junction

allowing current to flow under forward bias conditions but allowing very little current to flow in the opposite direction.

Unlike the thermionic diode, the pn diode needs no heater supply. On the debit side, the semiconductor rectifier cannot be operated at temperatures very much above room temperature and cannot withstand high voltages in the reverse bias condition. It is very much a low-voltage device.

The general symbol for a rectifier (either valve or semiconductor) is shown in Fig. 6.7.

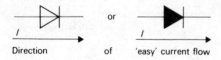

Fig. 6.7 Symbol for a rectifier

Rectification using thermionic diodes

An alternating voltage can be converted to a direct voltage by a thermionic diode using the circuit shown in Fig. 6.8. The low-tension supply is provided from a suitable source and the alternating voltage to be rectified is connected in series with the diode and a resistor. Current flows through the diode from anode to cathode only when the anode is positive to the cathode. Consequently, current flows only for half the cycle of the applied voltage. The current through the resistor consists of pulses separated by gaps where no current flows. The voltage produced across the resistor follows the same waveform and since only half waves are produced the circuit is called a half-wave rectifier. A more practical circuit is shown in Fig. 6.9.

Here the battery supply for the heater is removed and the low-tension supply is derived from the alternating supply by means of a suitable transformer. The transformer is an extremely useful device having the capability of readily changing alternating voltages from one value to another. The transformer consists basically of a closed laminated silicon-steel core and wound

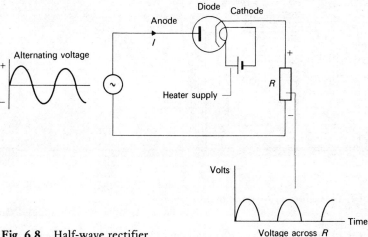

Fig. 6.8 Half-wave rectifier

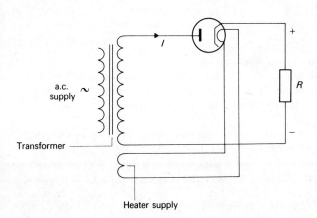

Fig. 6.9 Practical half-wave rectifier using a thermionic diode

on this core are two or more windings of copper wire. One winding, the primary, is connected to the a.c. supply, the other winding, the secondary, produces the required voltage. (See Fig. 6.10.) The ratio of the primary to the secondary voltage is the same as the ratio of the number of turns of wire on the primary to the number of turns of wire on the secondary. Mathematically

$$\frac{V_{\text{primary}}}{V_{\text{secondary}}} = \frac{N_{\text{primary}}}{N_{\text{secondary}}}$$

where V represents a voltage and N the number of turns of wire. If the primary has 800 turns and the secondary 200 turns, the secondary voltage is $\frac{1}{4}$ the value of the primary.

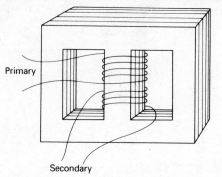

Fig. 6.10 Transformer (diagrammatic)

If the primary voltage is 240 V and there are 800 turns on the primary, and 200 V are needed on the secondary, then the number of secondary turns required is given by

$$\frac{240}{200} = \frac{800}{N_{secondary}}$$

From this, $N_{secondary}$ works out at 667 turns..
If the heater circuit for the diode requires 5 volts then an additional heater winding can be wound on the same transformer core. It can be shown, using the same formula as before, that about 17 turns are needed to provide the low-tension supply.

The symbol for a transformer is that used in Fig. 6.9. A primary winding and two secondary windings are shown.

The voltage produced by a half-wave rectifier is direct (i.e. always of one polarity) but is not steady. The situation can be improved by adding a reservoir capacitor across the load resistor. The action of this component is to charge up (i.e. 'fill up with electricity') during the first part of the positive voltage. When the voltage starts to fall, the energy contained in the reservoir capacitor now leaks through the load. The 'level' in the reservoir falls only slightly before the next cycle recharges it, i.e. fills it up again.

The final voltage now consists of a steady value with a superimposed 'ripple voltage' (Fig. 6.11). Further filter components can be introduced to reduce this 'ripple', for example, a series 'choke' can be added. The choke is rather similar in construction to a transformer which has no secondary winding. It has the ability of passing the direct component of the voltage from the rectifier easily but impeding the a.c. component — the ripple. The smoothing capacitor shown in Fig. 6.12 charges up to the steady d.c. value of the voltage but provides a by-pass for the a.c. component. The final ripple voltage across the load is usually less than 1 per cent of the steady voltage when a filter of this type is used. Further improvements can occur if the circuit is changed from a half-wave rectifier to a full-wave rectifier.

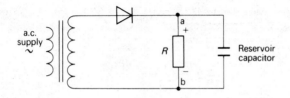

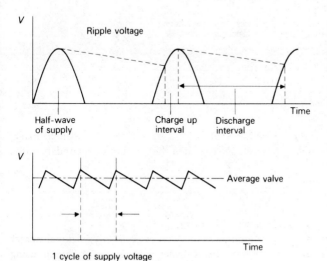

Fig. 6.11 Effect of reservoir capacitor

Figure 6.12 gives a practical example of this circuit.

Two diodes are required — often these are enclosed in one glass envelope with a common cathode. The secondary winding feeding the two anodes is in two halves so that the anodes are alternately positive and negative to the centre point of this winding. The result is that the two halves of the double diode conduct alternately and rectification occurs during the negative half-cycle as well as the positive half-cycle of the supply voltage. The effect is to fill up the gaps which are present in the voltage waveform of the half-wave rectifier. At the same time this reduces the magnitude of the ripple voltage and doubles the frequency, i.e. the ripple voltage repeats itself twice per cycle of the supply voltage. This results in a more efficient smoothing of the rectified voltage and a considerable reduction of the ripple voltage. In general the smaller the load current the smaller the ripple voltage. Increasing the size of the reservoir capacitor, i.e. making its electrical capacity larger, will reduce the ripple voltage but unfortunately there are limitations. If the capacity is too great the current demanded from the rectifier is very large during the recharging part of the cycle and damage can be caused to the valve. Another disadvantage is the fact that the physical size of a given capacitor increases as

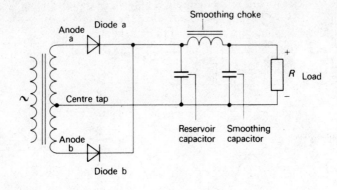

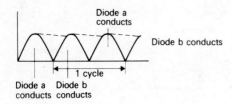

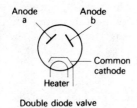

Fig. 6.12 Full-wave rectifier with smoothing components added

the electrical capacity increases and often there is only a limited amount of space available in electronic equipment. Further details on capacitors will be found in Chapter 7.

It should be remembered incidentally that the heaters of thermionic devices require a 'warm up' period, so that rectifier circuits do not immediately provide a direct voltage after 'switch on'.

Rectification using semiconductor diodes

Normally semiconductor diodes (pn junctions) can be used in place of thermionic valves if the voltages encountered are much lower. Semiconductors usually operate in the range 5 V to 20 V. The main advantage over the valve is the fact that no low-tension supply is needed. The semiconductor diode is usually much smaller in physical size than the valve. The full-wave rectifier employing semiconductors is identical to that shown in Fig. 6.12.

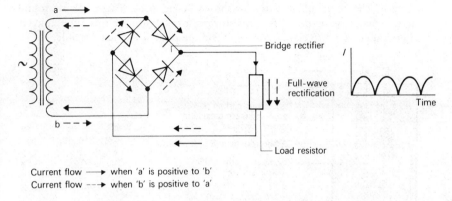

Current flow ⟶ when 'a' is positive to 'b'
Current flow ---▶ when 'b' is positive to 'a'

Fig. 6.13 Full-wave rectification using a bridge rectifier

An alternative form of rectifier is shown in Fig. 6.13, which dispenses with the need of a centre tapping point on the transformer. The circuit is called a bridge rectifier and the direction of current flow for the positive and negative half-cycles of the a.c. supply are shown. The rectified output is full-wave and a reservoir capacitor and a series choke with a further smoothing capacitor can be added in much the same way as the valve rectifier. The physical size of the capacitors is smaller than with the valve circuit as smaller voltages are being used.

Power supplies, general points

Comparing in general the thermionic valve and the semiconductor rectifier, the following points emerge:
1. Thermionic circuits operate at relatively high voltages, 100–300 V.
2. Semiconductor circuits operate at relatively low voltages, 5–20 V.
3. A low-tension supply is needed for thermionic valves but is not required for semiconductors.
4. The physical size of semiconductors is smaller than valves.
5. The physical size of capacitors used with semiconductor rectifiers is usually smaller than those employed with valves.
6. Both full-wave and half-wave rectification are available with either valves or semiconductors.

Of course it is possible to dispense with rectifiers if a suitable d.c. supply is available. Again semiconductors show a distinct advantage as only low voltages are needed. A typical semiconductor circuit employs a 9-V supply readily attainable from a small battery.

Batteries

These fall into two main types — non-rechargeable and rechargeable or primary and secondary cells. The former are usually of the 'Leclanché' pattern. The actual construction is shown in Fig. 6.14. These cells are found in the majority of portable transistor radios. Devices of this type convert chemical energy into electrical energy and the process is non-reversible.

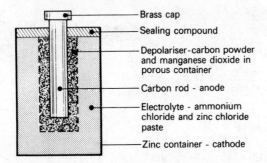

Fig. 6.14 Leclanché cell (dry)

Each cell of a battery produces 1·5 V so that a 9-V battery requires six cells arranged in series, that is, the positive pole of one cell (carbon rod) is connected to the negative pole (zinc container) of the adjacent cell. See Fig. 6.15.

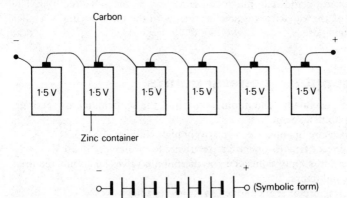

Fig. 6.15 Six 1·5-V cells in series

The life of a primary cell is limited for two reasons. Firstly, bubbles of hydrogen tend to form round the positive plate, which produces an insulating layer and prevents the further flow of electric current. This action is called polarisation. It is necessary to get rid of these bubbles. A depolarisation agent (manganese dioxide and powdered carbon in the case of the Leclanché cell) has to be employed. The amount of hydrogen which a depolariser can absorb is limited so the useful life of a cell is controlled largely by the amount of active depolariser present. As the battery nears the end of its useful life the effective internal resistance produced by polarisation increases and in consequence the terminal voltage on load decreases. The second effect which limits the life of a cell is caused by impurities in the zinc, and is known as local action. Small internal cells are produced which generate localised voltages and currents and these eat away the zinc. This effect occurs whether or not the cell is connected to any circuit and thus the 'shelf life', i.e. the life in storage, is considerably affected by local action. Higher purity zinc is the main answer to this problem. The electrolyte which is basically responsible for the conversion of chemical energy into electrical energy is ammonium chloride (commonly referred to as sal-ammoniac). The current which a Leclanché cell can deliver is limited mainly by the volume of the depolariser. Cells capable of delivering fairly large currents (for short periods only) are therefore physically large.

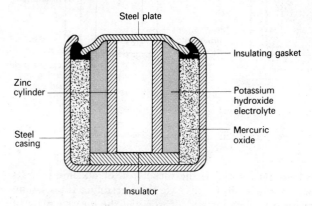

Fig. 6.16 A mercury cell

A relatively new form of primary cell is the mercury cell illustrated in Fig. 6.16. The steel container forms the positive electrode. This type of cell has the advantage of: (i) no polarisation under normal conditions; and (ii) no local action. It therefore produces a constant voltage — even when approaching the end of its useful life — this voltage is between 1·2 V and 1·3 V. It has a very long 'shelf life'. It is also size for size capable of producing a greater amount of electrical energy than its Leclanché counterpart. Unfortunately, largely because it employs more expensive chemicals — especially the mercuric oxide — it is more expensive to produce.

Secondary cells, unlike the primary cells just described, are rechargeable, i.e. it is possible to reverse the action and put into the cell electrical energy and convert it to chemical energy. The most common type is the lead-acid 'accumulator'. It is capable of delivering much higher currents than the Leclanché cell.

In a charged condition the positive plate is composed of lead peroxide and the negative plate is metallic lead. When discharged both plates turn to lead sulphate and the specific gravity of the electrolyte (sulphuric acid) falls. See Fig. 6.17. When fully charged the terminal voltage is 2·15 V and the specific gravity of the electrolyte is 1·205. When discharged these figures fall to 1·8 V and 1·185 respectively.

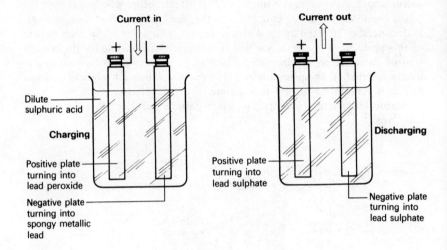

Fig. 6.17 A lead-acid secondary cell

Another form of secondary cell is the nickel-iron or alkaline cell. Its construction is shown in Fig. 6.18. In this case the electrolyte is potassium hydroxide. The negative plate is composed of iron and the positive plate of nickel hydroxide. Sometimes cadmium is added to the iron. This type of cell can withstand heavier currents during both charge and discharge but has a lower terminal voltage (1·2 V) and is more expensive to produce.

Similar cells (whether primary or secondary) may be connected in parallel (i.e. positive poles connected together and negative poles connected together). See Fig. 6.19. In this configuration the total current delivered by the battery is equally divided between the cells, e.g. three Leclanché cells in parallel delivering a total of 1·5 A, each deliver 0·5 A. The terminal voltage of cells in parallel is the same as that of a single cell.

If higher voltages are needed then cells must be connected in series and the current through any one cell is the same as the load current. Batteries can be made up of series and parallel combinations of cells if both high voltages and large currents are desired simultaneously. It should be pointed out that only identical cells should be connected in parallel. If for example a charged lead-acid cell is connected in parallel with a discharged lead-acid cell, current will circulate between the two, discharging one and charging the other.

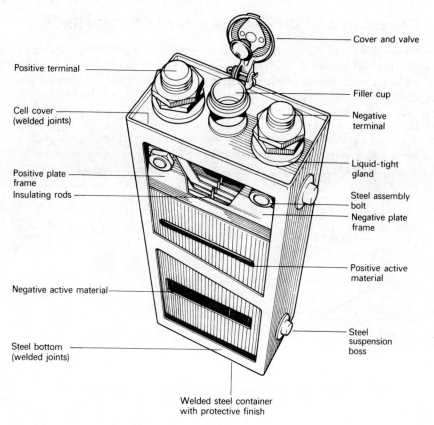

Fig. 6.18 Alkaline cell

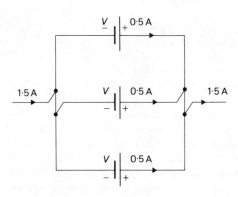

Fig. 6.19 Cells in parallel

Charging and discharging secondary cells

When discharging, i.e. delivering electrical energy to the circuit connected to the battery terminals, the terminal voltage falls. The graph showing this variation with time is illustrated in Fig. 6.20. The voltage remains fairly constant for about half the discharge time in the case of the lead-acid cell if the current demanded from the battery is not excessive. The basic unit of electrical charge is the **coulomb** (or ampere second), i.e. 1 ampere flowing for 1 second, or 2 amperes for $\frac{1}{2}$ second, etc. For secondary cells this is a rather small unit and the charge is often measured in the much larger unit of the ampere hour (Ah):

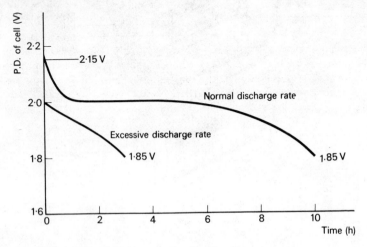

Fig. 6.20 Characteristics of a lead-acid cell

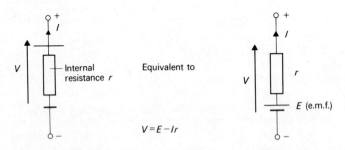

Fig. 6.21 Effect of internal resistance

1 Ah = 3600 coulombs, i.e. 1 ampere flowing for 3600 seconds.
When in a discharged state the internal resistance, r, increases causing the terminal voltage to fall. The voltage available when no current flows is called the electromotive force or e.m.f. E.

The terminal voltage V when current flows is given by $V = E - Ir$. See Fig. 6.21. Initially the value of r is considerably less than 1 ohm in a secondary cell (see Chapter 3).

When charging a secondary cell a direct voltage must be applied which is greater than the e.m.f. The positive pole of the battery must be connected to the positve terminal of the supply so that current is forced round the circuit in the opposite direction to that when discharging. The charging current can be controlled by altering the size of the supply voltage. The ampere hours during discharge are always less than those during charge since a certain amount of useful energy is lost during the charging process.

Amplifiers

Amplifiers employ triodes (thermionic three-electrode devices) or transistors (semiconductor three-electrode devices). Triodes are becoming less frequently used in modern electronic equipment. Transistors too have changed over the years. In their original commercial form they were in construction similar to that shown in Fig. 6.22. These are called bipolar transistors and although still

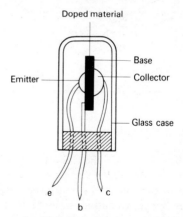

Fig. 6.22 Bipolar junction transistor

extensively used the more modern form is the planar construction (Fig. 6.23). In either case three electrodes are employed, the emitter, the base and the collector. The device can use pnp materials for the emitter, base and collector respectively, or npn materials. In each case the base (the 'meat' in the sandwich) is made as thin as possible. The symbols for the two transistors pnp and npn, whether silicon or germanium, are shown in Fig. 6.24. In normal use as an amplifier the emitter/base junction (pn or np) is forward biased and the base/collector junction is reverse biased. Remember that the term 'forward bias' implies that the polarity of the applied voltage is opposed to the potential barrier already existing across the pn junction. Under these conditions the pn junction is readily conducting. 'Reverse bias' means that the polarity of the applied voltage is in the same direction as that of the potential

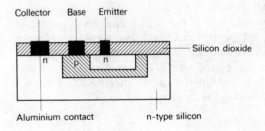

Fig. 6.23 Planar type transistor

barrier. The junction is virtually non-conducting in this state. A reversal of polarities of the supplies employed for pnp transistors is required to that of pnp. See Fig. 6.25.

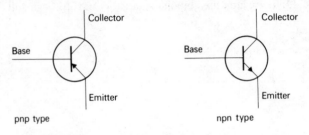

Fig. 6.24 Transistor symbols

The amplifier circuit uses a resistor in series with the collector and considering the case of the npn transistor, the flow of electrons between emitter and collector is controlled by the base current. With a pnp transistor the base current controls the flow of holes between emitter and collector. Note the polarities and directions of current flow in each case (see Fig. 6.25).

Small variations in base current cause significantly larger changes in collector current (typically a hundred times) and this provides the basis of amplification. Small changes in base/emitter voltages cause the base current to change developing in turn a much bigger change in collector current and consequently a much larger voltage change across the collector resistor.

The steady base-bias current is normally in the region of 100–200 μA (microamperes) and the steady collector current is around 3 or 4 mA (milliamperes).

The base/emitter bias battery can be dispensed with by using a resistor network (Fig. 6.26). Normally the bias voltage exceeds 1 volt.

Several amplifiers can be connected in cascade to increase the overall amplification, each being termed a 'stage'. Once two stages of amplification

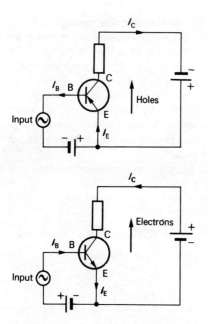

Fig. 6.25 pnp and npn amplifiers

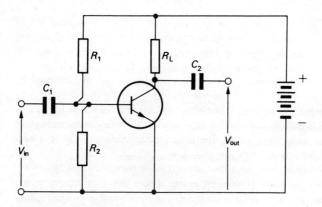

Fig. 6.26 Dispensing with bias battery

are reached, precautions must be taken to prevent instability. Each stage of amplification results in the reversal in phase of the input voltage so that with two stages the output and input are reversed twice and are now once more in phase. There is always a possibility of some of the output inadvertently reaching the input. If only 0·1 per cent of the output does this and the overall amplification exceeds 1000 times, then the system is liable to burst into oscillations. This is called positive feedback and is similar to the situation of 'howl back' when the output of a loudspeaker reaches the microphone in a public-address system.

Many semiconductor circuits today employ a technique in which many hundreds or even thousands of transistors are contained in a very small area indeed. Such circuits are called integrated circuits. Integrated circuits may have several stages of amplification but are carefully designed to prevent the sort of instability referred to. Amplifiers of the integrated circuit design frequently have two inputs marked + and − which are non-inverting and inverting respectively. By this is meant that in one case the output and input are in phase, in the other they are out of phase.

Any form of amplifier requires at least a supply voltage (around 9 V normally), an input and output.

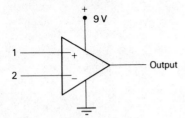

Fig. 6.27   Amplifier with both inverting and non-inverting inputs

The requirements are shown symbolically in Fig. 6.27. Frequently it is unnecessary to know in detail the actual construction of the circuit but merely to know where to connect the input and output and the supply battery. The main feature of the amplifier is its gain (i.e. output voltage ÷ input voltage). This can frequently be very high (around one million times). The other feature which must be known is the frequency range for which the amplifier is designed. Direct current amplifiers can amplify direct voltages but rarely operate satisfactorily at frequencies much above a few hundred hertz. Alternating current amplifiers have a frequency range extending from a few hundred hertz to kilohertz or even megahertz.

In a 'black box' approach amplifier internal constructional details are less important than the gain and frequency range. There are two other factors which must also be known about an amplifier: (1) the input current taken when an input voltage is applied − the ratio of input voltage to input current is known as the input impedance; and (2) the effective impedance offered at the output side of the amplifier. This is rather similar to the internal resistance of a cell and affects the size of the output of the amplifier when the output current is fed into another circuit. This output impedance becomes of greater significance when dealing with power amplifiers.

Field-effect transistors (another form of transistor) are generally characterised by a very much higher input impedance (the ratio of input voltage to input current) than the bipolar transistor. So much so that virtually no input current flows when an input voltage is applied. In these devices the 'base'

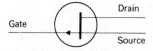
p-channel junction F.E.T.

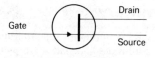
n-channel junction F.E.T.

Fig. 6.28 F.E.T. symbols

electrode is referred to as the gate, the emitter electrode is the source and the collector electrode the drain. While the action of the field-effect transistor is significantly different from the bipolar transistor the amplifying action is somewhat similar, except that the gate/source voltage is the controlling factor rather than the base current. The symbol for a field-effect transistor (usually referred to as an F.E.T.) is shown in Fig. 6.28. Most amplifiers are concerned with the amplification of voltage and the voltage gain (output voltage ÷ input voltage) is the main consideration in choosing an amplifier. Some amplifiers are designed to provide power rather than voltage and these tend to be physically rather large.

The power gain of an amplifier is defined as

$$\text{power gain} = \frac{\text{output power}}{\text{input power}}$$

In many cases power is given by the product of current and voltage even when alternating conditions are considered.

For such cases the power gain is given by

$$\text{power gain} = \frac{\text{output current} \times \text{output voltage}}{\text{input current} \times \text{input voltage}}$$

(This occurs again in Chapter 7.)

Amplifiers are also designed to handle only limited input voltages and power. If these limits are exceeded then the output of the amplifier is no longer a reasonable replica of the input. Under these conditions distortion occurs.

When selecting an amplifier therefore it is necessary to know:
1. whether a power or a voltage amplifier is required;
2. the gain of the amplifier;
3. the frequency range of the amplifier;
4. the maximum input it can handle;
5. the input impedance;
6. the output impedance.

Other factors which may affect the choice include the size and cost.

Nowadays increasingly one tends to purchase ready-made amplifiers rather than designing and building them. Such amplifiers are frequently of the integrated circuit type.

Nevertheless one must remember that there are occasions when one has to design amplifiers and one inevitably encounters from time to time older equipment employing older-type transistor circuits or even thermionic valves.

Summary

Principle of action of the thermionic diode and the pn junction. Rectification. Full-wave and half-wave rectification and the smoothing of the rectified wave. Batteries, primary and secondary types. Charging and discharging – the ampere hour. The transistor amplifier. The significant requirements of an amplifier – bandwidth, gain, input and output impedance. Power gain.

Questions

1. What, briefly, are the advantages and disadvantages of the semiconductor rectifier over the thermionic diode? Include in your answer some reference to environmental conditions, e.g. heat, humidity, corrosive gases, etc.

2. What are the 'doping agents' used in: (i) germanium; (ii) silicon? Consider both p and n materials.

3. Find the size of the potential barrier which builds up across a pn junction for both germanium and silicon devices.

4. A transformer is required for a rectifier circuit using thermionic diodes. The primary, which is to be connected to the 230-V, 50-Hz supply, has 1000 turns. Three secondary windings are required giving: (a) 450 V centre tapped; (b) 5 V low tension; (c) 6·3 V low tension.
   How many secondary turns are needed?
   Ans. 1956 t, 22 t, 27 t.

5. A full-wave rectifier produces a peak output from a 50-Hz mains supply of 150 V. On load with a smoothing circuit included the voltage falls to 140 V before the reservoir capacitor recharges. Draw the waveform of the output voltage on load and calculate: (i) the average value of the direct voltage output; (ii) the frequency of the ripple voltage; (iii) the time interval between the voltage peaks of the ripple.
   Ans. 145 V, 100 Hz, 0·01 s.

6. A secondary cell on charge takes an average current of 1·5 A for 24 hours. On discharge it produces an average of 12 V across a 2·4-Ω resistor for 6 hours. Calculate its efficiency on an ampere-hour basis.
   Ans. 83·3 per cent.

7. The internal resistance of a battery is 0·2 Ω. On open circuit it gives 12·6 V. What is:
   1. the terminal voltage when delivering 10 A?
   2. the terminal voltage when connected to a 5-Ω resistance?
   3. the maximum current that this battery can theoretically supply?
   Ans. 10·6 V, 12·1 V, 63 A.

8. An amplifier has the following characteristics
   voltage gain 1500
   input resistance (impedance) 10 000 Ω

The output is connected to a 20 000-Ω load.
Find the output voltage and the output power when the input is 0·01 V (10 mV).
What is the power gain?
Ans. 15 V, 0·011 25 W, 1 125 000.

9. Each stage of a two-stage amplifier using identical components in each stage produces a voltage gain of 250. What is the smallest percentage of the output which when fed back can cause oscillation or 'howl back'?
Ans. 0·0016 per cent.

10. Without referring to the text, draw in symbolic form and with the right battery polarities a simple amplifier for: (i) a pnp transistor; (ii) an npn transistor.
Indicate on your diagram the base, the collector and the emitter.

11. Compare from the point of view of cost the following:
   1. a thermionic rectifier unit producing a smoothed direct output of 12 V, 1 A continuous;
   2. a rectifier unit using semiconductors but producing the same output;
   3. a primary battery producing the same output;
   4. a secondary battery producing the same output.
   Take into account initial costs and the operating costs over a six-month period.

Chapter 7

Components

Active and passive components

An electronic circuit contains elements or components which can be considered to fall into two main categories: (i) active components; (ii) passive components. The former usually require a supply of direct current obtained either from the electricity mains via a suitable rectifying network or from a battery before they can operate successfully. Passive components do not require a supply of electricity to function.

Thermionic valves and transistors fall naturally in the first category. Resistors, capacitors and inductors form the bulk of the second group. From this definition connectors, plugs, sockets and many switches are all passive components. It is a little more difficult to know in which category to place such items as electromagnetic relays, certain microphones and lamps, since one could argue that they also need an electrical supply before they can function. Nevertheless they are usually regarded as 'passive'. Up until the widespread use of transistors there was a fairly clear-cut division between active and passive components. Valves could be easily identified. Resistors and capacitors were so different in appearance from valves that no confusion could arise.

With transistors becoming so cheap the situation has become significantly more complicated since it is possible to cause transistors to act as resistors and even capacitors, as well as amplifying devices. So it is not always easy to say for certain merely by looking at a modern electronic circuit whether a component is acting in a passive or an active manner.

With integrated circuits the separation of active and passive components is even more difficult as often the components are totally enclosed in a small box and are anyway so small as to make identification virtually impossible without the aid of an optical magnifier. Most technicians use the following condition to help differentiate between the two.

Active components are normally capable of amplification whereas passive components are not.

Even this additional 'test' raises problems as logic circuits, which one would normally consider active, do not amplify in the normally accepted sense of the word, and the pn junction diode apparently falls into the passive category.

Most technicians and engineers normally regard diodes, valves, transistors and integrated circuits as 'active' and everything else as 'passive'.

Passive components – resistors

The three most common passive components are resistors, capacitors and inductors. Each behaves in a characteristic fashion and their functions will be described in turn. When talking about the 'size' we are normally considering their electrical magnitude rather than their physical bulk. Considering resistors first of all – resistors have the property of resistance as they 'resist' the flow of current. They have already been encountered in Chapter 3 and can be defined from one form of Ohm's law, namely

$$R = \frac{V}{I}$$

Thus the electrical size of a resistor is measured in ohms (Ω) and can be found by placing a known voltage across the component and measuring the current flowing.

If for example 18 V across a resistance R causes 2 A to flow then the resistance value is given by

$$R = \frac{18}{2} = 9 \; \Omega$$

Resistances can vary greatly in electrical value – from fractions of an ohm to many millions of ohms – and it is usual to express high values of resistance by the use of the terms kilo (k) meaning thousands of, and mega (M) meaning millions of.

Thus 1000 Ω (one thousand ohms) = 1 kΩ
    10 kΩ = 10 000 Ω
    1·68 kΩ = 1680 Ω and so on.

1 kΩ is usually termed a 'kilohm'.

In the same way

1 000 000 Ω (one million ohms) = 1 MΩ
10 MΩ = 10 000 000 Ω
2·47 MΩ = 2 470 000 Ω

It also follows that 1 MΩ = 1000 kΩ
1 MΩ is usually termed a 'megohm'.

The use of k and M can also be related to voltage and current. Thus
12 kV = 12 000 V
i.e. 12 kilovolts = 12 thousand volts.

Only rarely in electronic circuits do we deal with voltages of this magnitude.

When dealing with current we often have to express a current flow in an electronic circuit as a small fraction of an ampere.

We tend to use the terms milli (m) meaning one-thousandth ($\frac{1}{1000}$) and micro (μ) meaning one-millionth ($\frac{1}{1\,000\,000}$).

Thus 1000 mA (one thousand milliamperes) = 1 A (one ampere)

$$1\cdot 2 \text{ mA} = \frac{1\cdot 2}{1000} \text{ A or } 0\cdot 0012 \text{ A}.$$

In the same way
1 000 000 μA (one million microamperes) = 1 A

$$45 \,\mu\text{A} = \frac{45}{1\,000\,000} \text{ A or } 0\cdot 000\,045 \text{ A}.$$

It follows that 1000 μA = 1 mA.
Other prefixes can be used, e.g. 'nano' and 'pico'.

Table 7.1 Multiples

Symbol	Name	Meaning	Mathematical expression
k	kilo	a thousand	× 1000 or × 10^3
M	mega	a million	× 1 000 000 or × 10^6

Submultiples

m	milli	a thousandth	÷ 1000 or × 10^{-3}
μ	micro	a millionth	÷ 1 000 000 or × 10^{-6}
n	nano	a thousand-millionth	÷ 1 000 000 000 or × 10^{-9}
p	pico	a million-millionth	÷ 1 000 000 000 000 or × 10^{-12}

e.g. 12·7 kV = 12 700 V
12·7 mA = 0·0127 A
12·7 μF = 0·000 012 7 F

If current, voltage, or resistance are measured in these other units, care must be taken as Ohm's law is expressed in ohms, volts and amperes — not megohms, or kilovolts, or microamperes.

Thus if a voltage of 12 V causes a current of 3 mA to flow through a resistor the value of the resistance in ohms is

$$R = \frac{12}{0 \cdot 003} \quad \text{since 3 mA} = 0 \cdot 003 \text{ A}$$

i.e. $R = \dfrac{12\,000}{3} = 4000 \, \Omega \text{ or } 4 \text{ k}\Omega$

In the same way if a resistance of 150 kΩ is placed across a 25-V supply the current flowing is given by

$$I = \frac{V}{R} = \frac{25}{150\,000} \quad \text{since 150 k}\Omega = 150\,000 \, \Omega$$

Thus $I = 0 \cdot 000\,167$ A
or 167 μA

Power dissipated in resistors

The physical size of a resistor is not related in any way to its electrical value. Resistors of the value of several megohms may be physically smaller than resistors of only a few ohms. The factor that largely determines the physical size of the resistor is the maximum electrical power that it will have to handle. When current passes through a resistor it tends to get hot and its temperature rises. This heat has to be dissipated in the surrounding atmosphere. By making the resistor physically larger a bigger surface area is presented to the atmosphere for cooling. In some cases where large powers are involved, cooling fins are added to increase this surface area or a cooling fan drives cold air over the surface to keep the temperature down.

Electrical power in d.c. circuits is given simply by the product of current and voltage. The relationship is

power = current × voltage, or in symbols
$P = I \times V$

If the current is in amperes and the voltage in volts then the power is measured in **watts**. In many electronic circuits, especially those using transistors, powers seldom exceed a few watts and often the power in a particular component is measured in milliwatts (mW) — 1 mW being $\frac{1}{1000}$ of a watt.

In the same way as Ohm's law can be expressed in three ways, so can the power in a resistor.

Since $P = I \times V$ and $I = \dfrac{V}{R}$ from Ohm's law

then $P = \dfrac{V}{R} \times V$ or $\dfrac{V^2}{R}$.

In the same way, since $V = IR$ from Ohm's law

$$P = I \times IR \quad \text{or} \quad I^2 R.$$

The three formulae for power calculations are therefore:

(i) $P = IV$; (ii) $P = V^2/R$; (iii) $P = I^2 R$.

V, I and R must be expressed in the fundamental units (not subunits or multiples) if the power P is to be expressed in watts.

Thus a current of 0·15 A flowing though a resistor of value 20Ω dissipates a power P given by

$$\begin{aligned} P &= I^2 R \\ &= 0{\cdot}015^2 \times 20 \\ &= 0{\cdot}45 \text{ watt.} \end{aligned}$$

A current of 15 mA flowing through a resistor of value 20 kΩ dissipates a power P given by

$$\begin{aligned} P &= I^2 R \\ &= 0{\cdot}15^2 \times 20\ 000 \\ &= 4{\cdot}5 \text{ watts.} \end{aligned}$$

Power rating of resistors

Resistors have a 'power rating'. This is the maximum power which can be dissipated in a particular resistor without possible damage by heat. Small (physically speaking) resistors may be rated at $\frac{1}{4}$ watt. The value in ohms of a resistor and its power rating affect both the maximum current which can pass through it and the maximum voltage which can be applied across it.

Example 7.1

A 47-kΩ resistor is rated at $\frac{1}{4}$ watt. What is: (i) the maximum current which can be passed through it; and (ii) the maximum voltage which can be applied across it?

It is necessary to choose the most convenient power formulae of the three to be able to solve this problem.

From $P = I^2 R$

$$I^2 = \frac{P}{R} \text{ on dividing both sides by } R$$

$$\therefore \quad I = \sqrt{\frac{P}{R}} \text{ on taking the square root of both sides.}$$

Hence the maximum allowable current is

$$I = \sqrt{\frac{0{\cdot}25}{47\ 000}} = 0{\cdot}002\ 31 \text{ A}$$

hence $I = 2{\cdot}31$ mA

The voltage can be now found from the formula

$P = IV$

From this $V = \dfrac{P}{I}$ on dividing both sides by I.

Hence $V = \dfrac{0\cdot 25}{0\cdot 002\ 31} = 108$ V.

Resistors employed in electronic circuits are usually available in power ratings of ¼ watt, ½ watt, 1 watt and 2 watt. It is common practice to use the resistor having the lowest power rating consistent with the requirements of the circuit. Thus if it is known that the power in a resistor is not going to exceed 0·1 watt, then obviously a ¼-watt rating would be ample although a resistor of a higher rating (say ½ watt) could be used if a ¼-watt resistor was not available. If the power dissipation is liable to be 0·8 watt then a 1-watt resistor would have to be employed. In an amplifier circuit, usually ¼-watt resistors are used at the input end. At the output stage it is far more likely that resistors of a higher power rating are found to be necessary. Of course the value of the resistor (in ohms) is the first concern, not the power rating, i.e. the correct value of resistance is the first consideration and the power rating is then calculated.

Construction of resistors

Resistors can be manufactured in a variety of ways. One way normally employed with resistors of high power rating is to wind thin nichrome wire on a ceramic former. The thinner and the longer the wire the higher is its resistance. In order to make such resistors physically small, very fine-gauge wire is employed and the ends are then secured to tinned coated copper wires of much thicker gauge. Often the wire is then coated with some vitreous enamel. Different forms of wire-wound resistor are shown in Fig. 7.1. Such

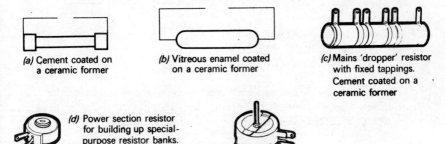

(a) Cement coated on a ceramic former

(b) Vitreous enamel coated on a ceramic former

(c) Mains 'dropper' resistor with fixed tappings. Cement coated on a ceramic former

(d) Power section resistor for building up special-purpose resistor banks. Cement coated on a ceramic former

(e) Mains 'dropper' resistor with adjustable tappings. Cement coated on a ceramic former

Fig. 7.1 Wire-wound resistors

resistors are called wire-wound resistors and often have a relatively high power rating (10 watts or more). As the power rating increases so must the physical size. When operating at their full rating the temperature of such resistors may well exceed 100 °C. Wire-wound resistors may be wound on an insulated card which is bent into the form of a circle. A conducting wiper may then bear on the resistor enabling a variable amount of resistance to be presented between one end of the resistance and the wiper arm. These are variable resistors. Usually such resistors have both ends of the wire-wound section brought out to connectors so that three connections are possible, one to each end and one to the wiper. These variable resistors are given the name potentiometer (Fig. 7.2).

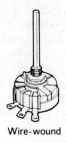

Wire-wound

Fig. 7.2 Variable resistor

The symbols for the various types of resistor are shown in Fig. 7.3. The move today is towards the block symbol rather than the zig-zag symbol.

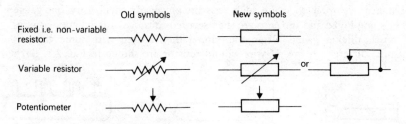

Fig. 7.3 Symbols used for resistors

Most non-variable resistors employed in electronic circuits are made from carbon mouldings or metal-oxide film. They are physically smaller than the wire-wound type but their rating is correspondingly smaller (up to 2 watts).

Resistors – colour coding

Typically, resistors are coded by means of colour bands which indicate the resistance values, but because they are made in very large numbers their values cannot be made very precise. They are graded in respect of a tolerance

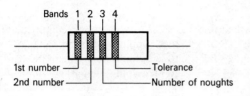

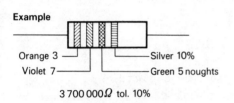

Fig. 7.4 Carbon resistors colour code

figure about an indicated value. This tolerance is either ±5%, ±10%, or ±20%. The tolerance figure is shown by means of a coloured band as well. The method of coding is shown in Fig. 7.4. Gold band means ±5%; Silver band means ±10%; no band means ±20%. Thus a resistor with the end band yellow, and second band violet, and the third band red is a 4700-Ω resistor. If the next band is silver this indicates a ±10 per cent tolerance and the actual value of the resistance can be anything between 4700 + 10% and 4700 − 10%

10 per cent of 4700 = 470
∴ the resistance could be between 5170 Ω and 4230 Ω.

The marked value is called the nominal value.
A resistor having a nominal value of 330 Ω and having no tolerance colour band could have a value between
330 ± 20%, i.e. between $330 + \frac{20}{100} \times 330 = 330 + 66$
and $330 - \frac{20}{100} \times 330 = 330 - 66$

The value could thus be anything between 396 Ω and 264 Ω.

A new British Standard is being introduced which uses a number and letter code to identify small resistors. Table 7.2 gives the details. In many respects it is of doubtful advantage over the colour code and since the letters

K and M are used with different meanings within the same code confusion might well arise.

Table 7.2 Number and letter code for resistors (B.S. 1852: 1967) (Courtesy R.S. Components Ltd)

The British Standard number and letter code for small resistors is now replacing the traditional colour code:

$0.47 \, \Omega$ would be marked R47 $100 \, \Omega$ would be marked 100R
$1 \, \Omega$ would be marked 1RO $1 \, k\Omega$ would be marked 1KO
$4.7 \, \Omega$ would be marked 4R7 $10 \, k\Omega$ would be marked 10K
$47 \, \Omega$ would be marked 47R $10 \, M\Omega$ would be marked 10M

After this code is added a letter to indicate tolerance:

$F = \pm 1\%$ $J = \pm 5\%$ $M = \pm 20\%$
$G = \pm 2\%$ $K = \pm 10\%$

Thus

R33M = $0.33 \, \Omega \pm 20\%$ 6K8K = $6.8 \, k\Omega \pm 10\%$
4R7K = $4.7 \, \Omega \pm 10\%$ 6K8M = $6.8 \, k\Omega \pm 20\%$
6K8F = $6.8 \, k\Omega \pm 1\%$ 68KK = $68 \, k\Omega \pm 10\%$
6K8G = $6.8 \, k\Omega \pm 2\%$ 4M7M = $4.7 \, M\Omega \pm 20\%$
6K8J = $6.8 \, k\Omega \pm 5\%$

Example 7.2

A resistor is marked
   1st band   Brown
   2nd band Black
   3rd band Orange
   No other band

What is this resistance and between what values does it lie?

   Brown (1) = 1   first unit
   Black   (0) = 0   second unit
   Orange (3) = 000   number of zeros
                     $10000 = 10 \, k\Omega$

Since no further band is given the tolerance is ±20 per cent. The resistance lies between $10\,000 + 2000$ and $10\,000 - 2000$, i.e. $12 \, k\Omega$ and $8 \, k\Omega$.

Example 7.3

The 1st band on a resistor is red, the 2nd band is red, the 3rd band is black, the 4th band is gold. What is the maximum value of the resistance?

   Red   = 2
   Red   = 2
   Black = __ no zeros
            22

Gold is ±5 per cent.
5 per cent of $22 \, \Omega$ is $1.1 \, \Omega$. The maximum value is $23.1 \, \Omega$.

The actual value may of course be less but the marking indicates that this resistance should not have a value exceeding 23·1 Ω.

Passive components – capacitors

These differ significantly both in construction and in action from resistors. In the first instance they do not provide a conducting path for electric current. They consist essentially of two metallic surfaces or plates separated by an insulator called the dielectric. The quality of a capacitor is governed by the insulating property of the dielectric. The electrical 'size' of a capacitor is referred to as the capacitance and this is affected by three factors: (i) the area of the plates; (ii) the distance they are apart; and (iii) the particular insulator used. See Fig. 7.5.

The capacitance increases with the area of the plates but decreases with the distance of their separation. Mathematically

$$C \text{ (capacitance)} \propto \frac{A \text{ (area of the plates)}}{d \text{ (distance of separation)}}$$

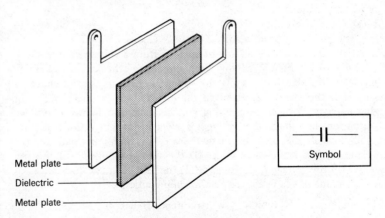

Fig. 7.5 A simple capacitor (exploded view)

The dielectric between the plates also increases the capacitance. Typical dielectric materials are paper, mica and many plastic materials.

Capacitors can be described as 'store houses' of electric charge. If a direct voltage is placed across the plates, the capacitor becomes charged, and when removed the voltage will be found still to be present across the plates. This will eventually leak away but capacitors of high capacitance using very good insulators as the dielectric will retain the electric charge for a considerable time. The charge on a capacitor is measured in coulombs (or ampere seconds) and is increased by either increasing the capacitance or increasing the potential difference applied across the plates (see Fig. 7.6). Mathematically

Q (charge) = C (capacitance) × V (voltage).

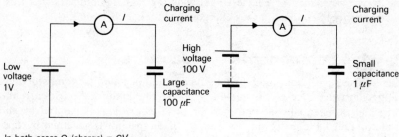

In both cases Q (charge) = CV
= 100 μC

Fig. 7.6 In both cases Q (charge) = CV = 100 μC

If the charge is in coulombs and the voltage in volts, then the capacitance is in **farads** (F). This is actually a very large unit and submultiples of much more commonly encountered. The microfarad (1 μF) is one millionth of a farad. An even smaller unit is employed especially when dealing with high-frequency circuits — this is the picofarad (1 pF). there are a million picofarads in one microfarad.

The capacitor not only stores electric charge; it also 'resists' changes in voltage, i.e. it does not respond readily to voltage changes.

A capacitor is charged by a flow of current from a voltage source and is discharged by allowing current to flow from it to some other circuit. Charging or discharging takes place only while the voltage across the plates is changing and the quicker the change takes place the greater will be the charging or discharging current. With an alternating voltage from the supply mains we have a situation in which the voltage is always changing. Consequently current will flow all the time although there is no conducting path for the current. The current I increases with: (i) the voltage applied; (ii) the frequency at which the voltage changes; and (iii) the value of the capacitance.

For the a.c. mains the following relationship applies.

I (current in amperes) = V (volts) × $2\pi f$ (frequency in hertz)
  × C (capacitance in farads)

Rearranging this formula

$$\frac{V}{I} = \frac{1}{2\pi fC}$$

The ratio $\frac{V}{I}$ is called the reactance (symbol X) and is measured in ohms.

It follows that X (reactance) = $\frac{1}{2\pi fC}$

The symbol for a capacitor is self evident (Fig. 7.6).

Example 7.4

What is the reactance of a 10 μF capacitance at 1 kHz?

$$X = \frac{1}{2\pi \times 1000 \times \frac{10}{1\,000\,000}} = \frac{1}{2\pi \times \frac{1}{100}} = \frac{100}{2\pi}$$

Hence $X = 15 \cdot 9\ \Omega$

Example 7.5

What current will a 5-μF capacitor pass when connected to the 240-V, 50-Hz supply?

The formula used is $I = V \times 2\pi f C$

hence $I = 240 \times 2 \times \pi \times 50 \times \frac{5}{1\,000\,000}$

$= 0 \cdot 377$ A

Construction of capacitors

One common method of construction is shown in Fig. 7.7. A sandwich consisting of alternate layers of thin metal foil and thin layers of waxed paper are rolled into a cylindrical 'swiss roll' so that relatively little volume is occupied for a relatively large capacitance. The arrangement is placed in a container and external electrical copper wires are fixed to the metal plates so that the capacitor may be connected into a circuit.

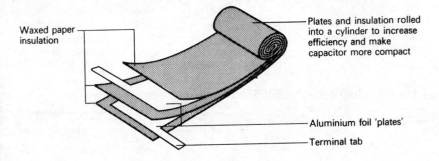

Fig. 7.7 Rolled paper dielectric capacitor

The physical size of a capacitor depends upon two factors: (1) the electrical capacitance; and (2) the maximum allowable voltage which is to be placed across the plates. The dielectric, being thin, is subjected to an electrical stress and is liable to break down, i.e. lose its insulating property. The insulator is actually punctured under such conditions and conducts electric current. This may lead to a sudden surge of current causing a mild explosion if the current is large. Capacitors have therefore a limited safe working voltage. The higher this voltage the larger the physical size of the capacitor, e.g. a 1-μF, 20-V working capacitor may be physically smaller than a 0·01-μF, 1000-V working capacitor. Mica capacitors are often employed in high-frequency circuits and are show in Fig. 7.8.

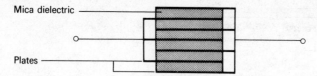

Fig. 7.8 Mica capacitor

Certain capacitors are of the electrolytic type and their construction is shown in Fig. 7.9. They rely for their operation on the deposition of an extremely thin film of aluminium oxide on metallic aluminium built up by the electrolytic action of aluminium electrodes in a dilute solution of boric acid (Fig. 7.10). In order to maintain this film a direct potential must always be placed across the plates of the capacitor of such a polarity as to continue the deposition process. If the polarity is reversed the film is removed and the capacitance reverts to a low value.

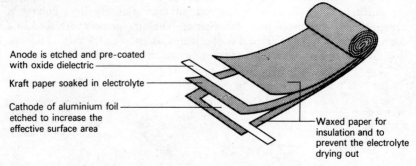

Fig. 7.9 Electrolytic capacitor

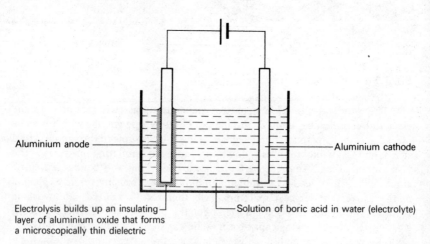

Fig. 7.10 Principle of the electrolytic capacitor

Tantalum is often preferred as a dielectric and can produce very high values of electrical capacitance in relatively small volumes, without resorting to electrolytic action. Such capacitors are unfortunately expensive.

Capacitors in series and parallel

Capacitors may be connected in series or parallel. With a series connection the same charging current flows through each. The charge, being a product of current and time, is therefore the same in each capacitor (Fig. 7.11). The voltage across each capacitor is different if the values of the capacitors are not the same.

Thus $V_{total} = V_1 + V_2 + V_3$

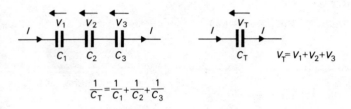

Fig. 7.11 Capacitors in series

But from the formula Q (charge) $= CV$

$$V_1 = \frac{Q}{C_1} \quad V_2 = \frac{Q}{C_2} \quad V_3 = \frac{Q}{V_3}$$

Thus $V_{total} = \dfrac{Q}{C_1} + \dfrac{Q}{C_2} + \dfrac{Q}{C_3}$

$$= Q\left(\frac{1}{C_1} + \frac{1}{C_2} + \frac{1}{C_3}\right)$$

If the three capacitors are replaced by a single equivalent capacitor C_T then the same voltage V_T must be produced by the same charge.

Hence $V_{total} = \dfrac{Q}{C_T}$

Comparing this result with the last equation

$$\frac{Q}{C_T} = Q\left(\frac{1}{C_1} + \frac{1}{C_2} + \frac{1}{C_3}\right)$$

or $\dfrac{1}{C_T} = \dfrac{1}{C_1} + \dfrac{1}{C_2} + \dfrac{1}{C_3}$

Example 7.6

What is the total capacitance of three $0{\cdot}1\text{-}\mu\text{F}$ capacitors connected in series?

$$\frac{1}{C_T} = \left(\frac{1}{0{\cdot}1} + \frac{1}{0{\cdot}1} + \frac{1}{0{\cdot}1}\right)$$

$$= \frac{3}{0{\cdot}1}$$

Inverting

$$C_T = \frac{0{\cdot}1}{3}\,\mu\text{F} = 0{\cdot}0333\,\mu\text{F}$$

Capacitors in parallel possess the same p.d. across their plates (Fig. 7.12). The charge on each is different if the capacitors have a different value.

Thus $Q_1 = C_1 V \quad Q_2 = C_2 V \quad Q_3 = C_3 V$

The total charge $Q_T = C_1 V + C_2 V + C_3 V$

$$= V(C_1 + C_2 + C_3)$$

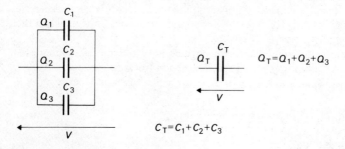

Fig. 7.12 Capacitors in parallel

If an equivalent capacitor C_T replaces the three capacitors the same voltage must produce the same charge

hence $Q_T = V C_T$

Comparing this with the last equation it follows that

$$C_T = C_1 + C_2 + C_3$$

Example 7.7

Three capacitors $0{\cdot}1\,\mu\text{F}$, $0{\cdot}2\,\mu\text{F}$, and $0{\cdot}5\,\mu\text{F}$ are connected: (a) in parallel; (b) in series. What is the total capacitance in each case?

(a) Parallel connection

$$C_T = C_1 + C_2 + C_3$$

Hence $C_T = 0{\cdot}1 + 0{\cdot}2 + 0{\cdot}5 = 0{\cdot}8\,\mu\text{F}$

(b) Series connection

$$\frac{1}{C_T} = \frac{1}{C_1} + \frac{1}{C_2} + \frac{1}{C_3}$$

$$\frac{1}{C_T} = \left(\frac{1}{0\cdot1} + \frac{1}{0\cdot2} + \frac{1}{0\cdot5}\right)$$

$$\frac{1}{C_T} = (10 + 5 + 2)$$

$$\frac{1}{C_T} = 17$$

$$\therefore C_T = \frac{1}{17}\,\mu F = 0\cdot0588\,\mu F$$

Variable capacitors

Since the capacitance depends upon the area of the plates, variable capacitors can be made by varying the area of intermesh of a number of plates of a capacitor. This can be achieved by pushing the plates together or rotating the vanes of a capacitor (Fig. 7.13). The minumum capacitance occurs when the intermesh is least.

The variation of capacitance with angular position can follow any prescribed pattern by altering the shape of the capacitor vanes.

The 'tuning' of a radio receiver is often achieved by the variation of a capacitor.

Figure 7.13 shows a number of different types of capacitor.

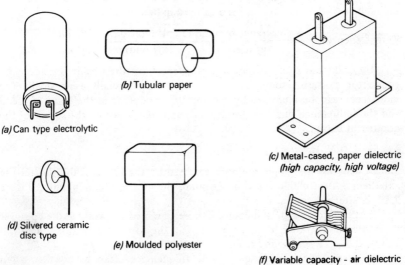

(a) Can type electrolytic
(b) Tubular paper
(c) Metal-cased, paper dielectric (high capacity, high voltage)
(d) Silvered ceramic disc type
(e) Moulded polyester
(f) Variable capacity - air dielectric

Fig. 7.13 Some typical capacitors

Passive components – inductors

Inductors, like resistors, allow the passage of electric current but unlike resistors which 'resist' current, inductors 'resist' changes in current. Thus while current remains unchanged inductors have no effect but as soon as the current changes from one value to another the inductor opposes the change and the amount of 'opposition' depends upon the rate at which the change takes place. For example, a change from 1 A to 2 A in current in 0·1 s represents a change rate of 1 A in 0·1 s or 10 A/s. If the same change occurred in 0·05 s then the change rate is doubled to 20 A/s and the inductor would have twice the effect.

When the current is alternating, as it does from the domestic supply, its value is always changing although its rate of change is not constant, so the presence of an inductor will always have an effect on the circuit.

Inductors, like capacitors, possess a quality called reactance when passing alternating current.

Reactance in this case is measured in ohms and is given by the ratio of voltage to current. Mathematically, with a sinewave of current the following relationship can be used.

Reactance (Symbol X) in ohms = $\dfrac{\text{voltage }(V)}{\text{current }(I)}$

In this sense the relationship is similar to the case of capacitance.

In order, however, to take into account the rate of change of the current the reactance is related to the frequency of the supply as follows:

X (ohms) = $2\pi f L$

f is the frequency in cycles per second or hertz of the current. L is called the inductance of the inductor and is measured in **henrys** (H).

Thus L (henrys) = $\dfrac{X \text{ (reactance in ohms)}}{2\pi f \text{ (hertz)}} = \dfrac{V \text{ (voltage)}}{2\pi f I \text{ (current)}}$

The inductance of an inductor is affected by the way it is made. Any conductor possesses some inductance but this is usually very small. The inductance can be increased by shaping the conductor in the form of a coil and the inductance furthermore depends approximately on the square of the number of turns N

i.e. $L \propto N^2$ approximately.

In general, the larger the area enclosed by the coil and the shorter the axial length over which the coil is wound, the greater the inductance. Figure 7.14 illustrates this.

Further increase in inductance can be obtained by winding the coil over an iron core. The core if closed on itself in the form of a ring or square will still further increase the inductance (Fig. 7.15).

Inductors which are used at low frequencies often have a laminated silicon-steel closed core – this type of construction being necessary to reduce

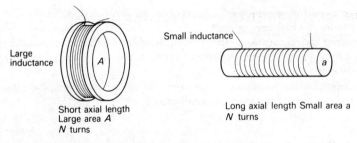

Fig. 7.14 Effect of dimensions on inductance

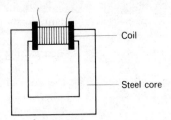

Fig. 7.15 Choke (large inductance)

losses. A winding of a transformer can be used as a low-frequency inductor. Low-frequency iron-cored inductors are frequently termed chokes. At higher frequencies ferromagnetic dust in a ceramic bonding is used as the core.

Inductors are normally available as standard items only when dealing with fairly large values of inductance (1 H upwards). Small inductors can be easily made by winding the necessary turns of wire on a 'former' (Fig. 7.16). The design of an inductor is influenced by the current which it will have to carry. Thus a 2-H inductor which has to carry 1 A will require thicker gauge wire than an inductor of the same inductance but needing to pass only 0·01 A, and will therefore be physically larger.

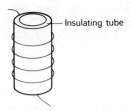

Fig. 7.16 Small inductance

The two factors which affect the choice of an inductor are therefore inductance and current-carrying capacity. It should be realised that there is a certain amount of resistance inherently associated with any inductor due to the resistance of the conducting wire used in its construction. Finer wire naturally produces a higher resistance.

Inductance boxes are sometimes used whereby inductances may be switched into a circuit by known amounts; they are rather similar to the decade resistance boxes.

Inductors in series behave similarly to resistors in series, that is, the total inductance is the sum of the individual inductances.

i.e. $L_T = L_1 + L_2 + L_3$

The symbols for inductors are shown in Fig. 7.17.

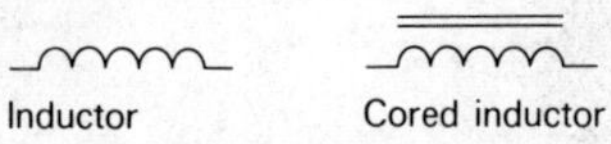

Inductor   Cored inductor

Fig. 7.17 Symbols for inductors

Example 7.8

A pure inductor (i.e. one possessing negligible resistance) is connected to the 230-V, 50-Hz domestic supply. A current of 2 A flows. What is the value of the inductance?

Reactance $X = \dfrac{\text{voltage}}{\text{current}} = \dfrac{230}{2} = 115 \ \Omega$

Inductance $L = \dfrac{X}{2\pi f} = \dfrac{115}{2 \times 3 \cdot 14 \times 50} = 0 \cdot 366$ H.

Example 7.9

If the inductor in the previous example has 100 turns of wire and another coil is now wound of 150 turns but otherwise having identical dimensions, what current would it pass if connected to the same supply?

Inductance $L \propto N^2$
If $L_1 = 0 \cdot 366$ H and has 100 turns
$0 \cdot 366 \propto 100^2$
Similarly $L_2 \propto 150^2$
Hence $\dfrac{L_2}{L_1} = \dfrac{150^2}{100^2} = 2 \cdot 25$
Therefore
$L_2 = 0 \cdot 366 \times 2 \cdot 25 = 0 \cdot 824$ H
Reactance
$X_2 = 0 \cdot 824 \times 2\pi \times 50 = 250 \ \Omega$
Hence $I = \dfrac{V}{X_2} = 0 \cdot 888$ A

Variable inductors

Inductance of a coil is dependent upon the physical construction of the coil, e.g. the number of turns, the diameter of each turn and the axial length over which it is wound. It is also affected by the presence of a ferromagnetic core.

The inductance is increased by the ferromagnetic material so that if the core is arranged to move into the core along its axis the inductance is increased (Fig. 7.18). The coil in some radio receivers is 'tuned' by this method, i.e. its value is altered.

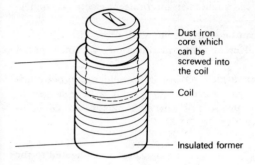

Fig. 7.18 Variable inductor

The three basic passive components — resistors, inductors and capacitors — have certain similarities and these are summarised in Table 7.3.

Table 7.3

Component	Symbol	Unit ratio V/I	Effect	Rating
Resistor	R	(ohm) Ω	'resists' current	Power
Capacitor	C	(farad) F X (in Ω) = $\frac{1}{2\pi fC}$	'resists' change in voltage	Voltage
Inductor	L	(henry) H X (in Ω) = $2\pi fL$	'resists' change in current	Current

Series connection	Parallel connection
$R_T = R_1 + R_2 + R_3$	$\frac{1}{R_T} = \frac{1}{R_1} + \frac{1}{R_2} + \frac{1}{R_3}$
$\frac{1}{C_T} = \frac{1}{C_1} + \frac{1}{C_2} + \frac{1}{C_3}$	$C_T = C_1 + C_2 + C_3$
$L_T = L_1 + L_2 + L_3$	

120 Connections, connectors and interconnectors

Components in an electronic circuit in which discrete resistors, capacitors, inductors, transistors, etc. are used are normally connected together by soldering the tag ends of the components either to posts or tags on an insulated base board or to the appropriate points of a printed circuit board.

Solder is a tin/lead alloy which has a comparatively low melting point. A heated copper soldering 'iron' is used to melt the solder and a resin flux (often contained as a core within the solder) allows the molten metal to flow freely on clean copper surfaces. If the surface is dirty a poor electrical joint (termed a dry joint) results, which produces an intermittent electrical connection.

Metals other than copper can be soldered but with some difficulty. It is often necessary to employ a different flux and sometimes a rather special soldering technique.

As the components in a circuit may get hot in the soldering process it is necessary to ensure that the soldering is carried out quickly. Where damage to the component may ensue then some sort of 'heat sink' must be employed to conduct the unwanted heat away. See Fig. 7.19.

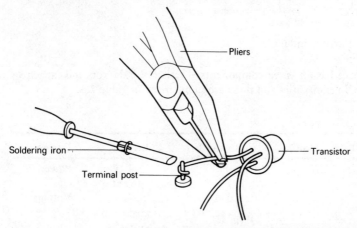

Fig. 7.19 Showing the 'heat sink' provided by the pliers conducting the unwanted heat from joint and thus keeping the transistor cool

Connectors are used to interconnect electronic subsystems, e.g. printed circuit boards. Frequently these are of the plug and socket type. Typically with printed boards the parts of the circuit which have to be connected are taken out to copper edge connectors which fit into the sockets (Fig. 7.20). Flexible wires are sometimes attached to the socket connectors and the other ends are soldered to external systems or parts of systems. Often the sockets themselves are fitted to a rack so that the printed boards may be easily removed. In this case there is no need to have a flexible interconnector. Where parts of a system are physically separated by several metres, a flat flexible connector as shown in Fig. 7.21 is employed.

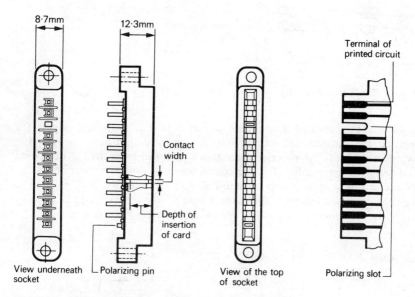

Fig. 7.20 Plug and sockets for printed circuit boards

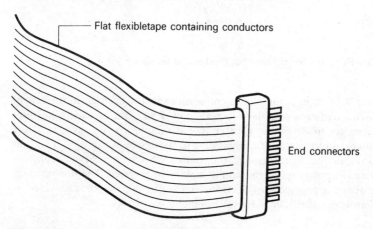

Fig. 7.21 Flexible flat interconnectors

Connections to instruments and test gear are usually made by either 'spade' terminals (flat notched-end connectors which fit under a screwed terminal) or a plug (banana plug) which slots into a round hole. Temporary connections are made using 'crocodile' clips. See Fig. 7.22.

Active components

Two main categories of active components exist: (i) thermionic devices; (ii)

Fig. 7.22 'Crocodile' clip

semiconductor devices. Thermionic devices require a low-tension supply for heating the cathode and can be directly heated or indirectly heated. The symbols used for the two-electrode device (a diode) have already been used but are given again in Fig. 7.23 for completeness. A three-electrode device is a triode and its symbol is also given in Fig. 7.23. The electrodes are referred to as cathode, anode and grid. The diode is used mainly for rectifying alternating currents; the triode is used mainly for the amplification of voltages.

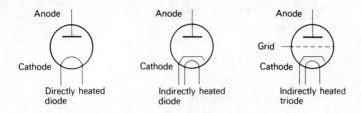

Fig. 7.23 Symbols for thermionic diodes and triodes

Figure 7.24 depicts the equivalent semiconductor devices. In the case of the transistor either a pnp device or an npn device can be used. The three connections are made to the emitter (corresponding to the cathode of the triode valve), the collector (corresponding to the anode of the triode) and the base (corresponding to the grid of the triode). When an npn device replaces a pnp device the polarities of the supply have to be reversed. The flow through the npn device is provided by electrons and in the pnp device is provided by holes. Transistors of this type are called bipolar transistors.

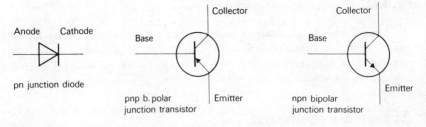

Fig. 7.24 Semiconductor active circuit symbols

New inventions called field-effect transistors (F.E.T.s) often replace the transistor just described. They behave rather similarly to the triode valve but are actually semiconductor devices. The emitter is renamed the source, the collector becomes the drain and the base is now the gate. Figure 7.25 shows the symbol for an F.E.T. There are now variations of this particular device which merely illustrate once more the rapid changes occurring in electronics.

p-channel junction F.E.T.   n-channel junction F.E.T.

Fig. 7.25   F.E.T. symbols

Because of the smallness of many of the active components it is possible to place a number of such units very close to each other and produce a complete amplifier in an integrated circuit, and we then revert to the block diagram approach used in Chapter 4.

Logic units

These are becoming of increasing importance so a little space will be spent in explaining what is meant by logic in both everyday as well as electronic terms.

Logic circuits can be employed to provide solutions to certain types of problems — especially mathematical ones — but they also feature in a number of everyday devices. Modern automatic washing machines use logic circuits. Needless to say this field of· electronics is finding increasing application to telecommunications.

Problems and their reduction to logic statements

Being presented with a problem is a situation with which we are all familiar. The problem may be a mathematical one which is capable of exact solution and may have only one correct answer, e.g. solve $2x + 17 = 4$.

Everyday problems are seldom of this type. There is frequently no completely correct solution since what is right or wrong may be a matter of opinion. Even what is considered by the majority as the correct solution today may be judged incorrect in a month's time in the light of additional information.

Often the problem is very complex and there is no easy route to any solution. For example, the problem generated by an economic crisis is one which appears to defy solution by successive governments.

Nevertheless any problem, however complicated, can be broken down into smaller problems or steps and often the alternative solutions to these more elemental problems are limited. Complex problems can sometimes be solved in this way. It is possible in many instances to so frame questions within a problem that only two alternatives are possible, an answer 'Yes' or an answer 'No', and the 'Don't know' answer is literally 'out of the question'.

Take a very trivial problem. 'What am I going to do tonight as I appear to have some free time?' Perhaps an unlikely question to many students but one which demands an answer.

The answer might be, 'I think I will visit the "local",' but that in turn may depend upon the weather and financial resources (although one could still visit the pub in the fond hope that some kind friend will buy one a drink). The problem can now be split down so that the alternatives are more clearly seen, e.g. 'I will go to the pub if it doesn't rain and I will buy a drink if my pay packet arrives in time'. Progress is now being made, the problem is being identified and the individual is being forced to take a decision.

The questions now are:
(i) Is it raining? Yes or No.
(ii) Has my pay packet arrived? Yes or No.

The decision to go to the pub rests on (i) and the decision to buy a drink rests on (ii). There is now a sequence in decision taking. This can be shown graphically by means of a diagram called an algorithm (Fig. 7.26). The clear cut answers Yes or No suggest a binary device, corresponding to two clearly defined states ON or OFF. (See Chapter 5.)

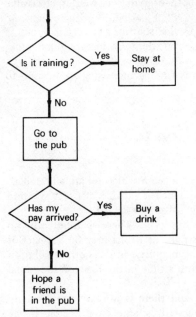

Fig. 7.26 An algorithm

In modern technology this is a situation which occurs time and time again and use is made of logic circuits to enable problems of this nature to be presented and solved. They are used extensively in computer circuits.

The four basic logic statements

A logic circuit is essentially a two-position device having states corresponding to ON and OFF, and behaves as a switch. The two states or positions of the logic circuit can therefore correspond to the answers Yes and No. It is a binary device and the alternatives can be considered to be a one (1) or a zero (0), i.e. one corresponds to the ON position (Yes) and 0 to the OFF position (No).

Before we can apply logic circuits to a problem it is necessary to provide a logic statement or question which has two alternatives.

Fortunately with logic of this sort there are a very limited number of situations and possibly because of this most people find problems on logic relatively easy. (It should be pointed out that framing the problem into the right format is more than half-way to providing a solution.)

Let us consider the alternative logic statements.

a.
The OR statement
'I can obtain some additional cash if I sell either my camera or my cycle.'
This is a logic statement in which three possibilities exist — if I sell either my camera or my cycle or both, I can obtain more cash. This in logic terms is known as an OR statement.
We can represent the obtaining of cash as F. The selling of the camera is A and the selling of the cycle as B.
Thus F occurs when either A or B or both occur.
i.e. $F = A$ OR B
In logic symbols
$F = A + B$

The + sign is not an arithmetical symbol but means OR. We can represent the situation electrically by a simple circuit — two switches in parallel. The lamp lights when either or both of the switches are closed. F is the lamp when alight. A and B are the two switches (Fig. 7.27).

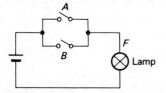

Fig. 7.27 The OR circuit

b.
The exclusive OR statement

The OR statement considered previously allowed three possibilities — one or the other or both. The exclusive OR statement allows only two.

'I am going to either the disco or the cinema but I cannot afford to go to both.'

One alternative excludes the other, there is no possibility of both. This is an exclusive OR statement and is represented by $F = A \oplus B$.

F represents the action of going somewhere, A is the action of visiting the disco, B the action of visiting the cinema.

An electrical system which is analogous to this is the two-way switch (Fig. 7.28). There is a clear alternative.

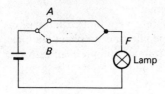

Fig. 7.28 The exclusive OR circuit

c.
The AND statement

'The river Thames flows under Waterloo Bridge and Tower Bridge.'
This is an AND statement. No alternative exists since the river must flow under both bridges.

This logic statement is represented by $F = A$ AND B or in symbols $F = A.B$ (the dot in this case means AND).

F represents the River Thames flowing, A represents flowing under the Waterloo Bridge, B under Tower Bridge. Both events A and B must occur if the River Thames flows at all.

An electrical circuit analogous to this situation is two switches in series. Both switches must be closed before the action F (the lamp lighting) occurs (Fig. 7.29).

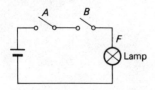

Fig. 7.29 The AND circuit

d.
The NOT statement
This is a situation in which the exact opposite happens to that in the statement.
'I am going out' is changed to 'I am not going out' by the addition of 'not'. The negation of an action is the NOT situation and can be expressed symbolically as follows:

$F = \bar{A}$ A means 'I am going out'
$\bar{A}$ means 'I am not going out'

Electrically speaking it is rather difficult to produce a simple NOT circuit. The nearest analogy is an inverter. A simple amplifier works essentially in this manner.

The algebra involving logic statements such as $F = A.B$ is called Boolean algebra and should not be confused with other algebraic methods. The four basic statements and the corresponding algebra can be built up to very complex situations involving a complicated sequence of decisions.

Logic symbols

Circuits which perform logic decisions are called logic gates and are available commercially. They usually require a small direct voltage (about 5 V) for operation. They can be made extremely small and therefore a large number can be packed into a very small space.

Standard symbols are employed for logic gates. It is perhaps unfortunate that the British standard and the American standard are not identical, so both are given. The number of inputs for AND and OR circuits can be increased to about 10, certain practical limitations prevent further increases. Many logic circuits have only two or three inputs.

Three- and two-input AND and OR symbols are shown in Fig. 7.30.
A NOT gate is illustrated in Fig. 7.31.

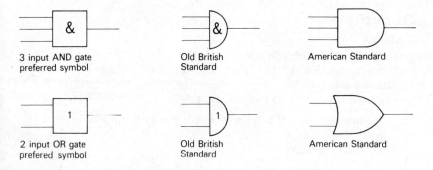

3 input AND gate preferred symbol   Old British Standard   American Standard

2 input OR gate prefered symbol   Old British Standard   American Standard

Fig. 7.30   AND and OR symbols

It is possible to have a NOT gate and an AND gate in series in a single unit. This is called a NAND gate and its symbol appears in Fig. 7.32.
Likewise a NOT and OR gate form a NOR circuit and this is illustrated in Fig. 7.33.

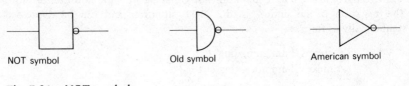

Fig. 7.31 NOT symbols

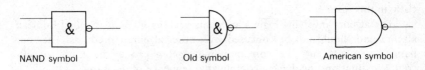

Fig. 7.32 NAND symbols

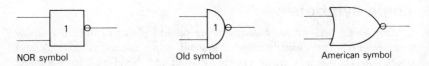

Fig. 7.33 NOR symbols

Summary

Circuit components can be considered passive or active. In the former lie the resistor, capacitor and inductor. Active circuits include transistors. One important section of active circuits involves logic gates.

Questions

Resistors
Select the answer which you think is correct.
1. A resistance of value 12 kΩ is placed across a 12-V supply. The current flowing is
   (a) 1 μA; (b) 1 mA; (c) 144 mA; (d) 1 A.

2. The current flowing through a resistor is 5 mA when placed across a 10-V supply. The value of the resistance is
   (a) 2 Ω; (b) 50 Ω; (c) 2 kΩ; (d) 50 kΩ.

3. The voltage across a 2-MΩ resistor when 2 μA flows is
   (a) 1 μV; (b) 1 mV; (c) 1 V; (d) 4 V.

4. A resistance has a value 1·47 kΩ. Its value in ohms is
   (a) 1470 Ω; (b) 14 700 Ω; (c) 147 000 Ω; (d) 1 470 000 Ω.
5. The power dissipated by a 1-kΩ resistor when placed across a 100-V supply is
   (a) 100 mW; (b) 1 W; (c) 10 W; (d) 100 W.
6. A resistor carries a current of 20 mA when connected to a 10-V supply. The power dissipated is
   (a) 2 mW; (b) 5 mW; (c) 0·2 W; (d) 2 W.
7. The maximum current which can be carried safely by a 0·25-MΩ, 1-W resistor is
   (a) 0·25 μA; (b) 0·25 mA; (c) 2 mA; (d) 4 mA.
8. A resistance has a nominal value of 1 kΩ and has a silver band indicating its tolerance. Its resistance lies between
   (a) 990 Ω and 1010 Ω; (b) 950 Ω and 1050 Ω; (c) 900 Ω and 1100 Ω; (d) 800 Ω and 1200 Ω.
9. A resistance has a nominal value of 330 Ω. Its actual value is 302 Ω. Should it have
   (a) a black band indicating its tolerance?
   (b) a gold band indicating its tolerance?
   (c) a silver band indicating its tolerance?
   (d) no band at all?
10. The colour code bands on a resistor starting from the end nearer the bands are brown, red, red. Its nominal resistance is
    (a) 122 Ω; (b) 1200 Ω; (c) 12 kΩ; (d) 10 MΩ.
11. The colour bands on a resistor starting from the end nearer the bands are brown, black, green. Its nominal resistance is
    (a) 106 Ω; (b) 16 kΩ; (c) 1 MΩ; (d) 10 MΩ.
12. The colour bands on a resistor starting from the end nearer the bands are yellow, violet, black. The nominal value of the resistance is
    (a) 47 Ω; (b) 470 Ω; (c) 4·7 kΩ; (d) 47 kΩ.

Capacitors

1. What is the charge in coulombs stored by a 0·2-μF capacitor when 250 V direct current are applied across it?
   Ans. 50 μC.
2. A 10·5-μF capacitor initially uncharged is charged at a steady rate of 0·2 mA for 5 s. What is the p.d. across its plates?
   Ans. 95·2 V.
3. Three 1-μF capacitors can be arranged in series, parallel or series-parallel arrangements. If the three capacitors are used in each case what total values of capacitance are available?
   Ans. 0·333 μF, 0·666 μF, 1·5 μF, 3 μF.

4. Four capacitors 1 μF, 250 V working; 2 μF, 100 V working; 5 μF, 150 V working; 10 μF, 50 V working are connected in series to a 500-V d.c. supply. What voltage exists across each capacitor? Is the circuit safe? What is the safe maximum voltage?
   Ans. 278 V, 139 V, 56 V, 28 V, 360 V.

5. A 50-V, 500-Hz a.c. supply is placed across a 0·01-μF capacitor. What alternating current flows? If the frequency were doubled and the voltage halved how would this affect the current?
   Ans. 25·1 μA.

6. A variable capacitor has movable vanes each of area 5 cm². When fully in mesh the capacitance is 540 pF. If the area of the intermeshed vanes is reduced to 2 cm², what is the new value of capacitance?
   Ans. 216 pF.

Inductors

1. A 240-V, 50-Hz a.c. supply is applied to an air-cored inductor of value 0·4 H. If the resistance of the coil can be neglected, what current flows? If a ferromagnetic core were now introduced how would this affect the current?
   Ans. 1·90 A.

2. A coil is marked 1·5 H, 250 mA rating 10-Ω resistance. If connected to: (i) a 15-V d.c. supply; (ii) a 15-V, 50-Hz supply would this be safe? Substantiate your answer in both cases.

3. Two coils of identical dimensions have respectively 250 turns and 350 turns. When the first is connected to a 1-Hz, 5-V supply, 10 mA flows. What current would flow if the second coil replaced the first? What is the inductance of each coil?
   Ans. 0·196 mA, 0·0796 H, 0·156 H.

4. Draw a graph showing how the current through a fixed coil varies with frequency if the magnitude of the voltage across it remains constant.

Active circuits

1. Without referring to the text draw the symbols for the following:
   (a) A semiconductor diode
   (b) A directly heated thermionic diode
   (c) An indirectly heated triode
   (d) An npn transistor
   (e) An F.E.T.
   (f) An AND gate
   (g) A NOR gate

2. I bought the following in a shop:
   2 kilos of sugar
   1 packet of teabags
   2 bottles of milk, in order to provide for 40 cups of tea.
   Is this an AND, NAND, OR, or NOR statement? Why?

3. The cicuit in Prob. Fig. 7.1 uses a number of switched relays. Show how logic gates could replace the switches.

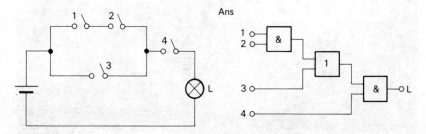

Prob. Fig. 7.1

4. The waveforms shown in Prob. Fig. 7.2 are applied to each of the two inputs to: (i) a two-input AND gate; (ii) a two-input OR gate. Draw the output in each case. (Positive logic is employed.)

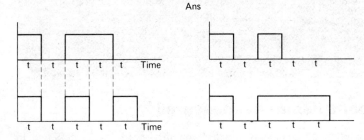

Prob. Fig. 7.2

Chapter 8

Measurements and instruments

The quantities to be measured

Three parameters or quantities which frequently need to be measured in electronic circuits are voltage, current and resistance. These quantities can be measured using three separate instruments called voltmeters, ammeters and ohmmeters respectively. In the case of the first two (voltmeters and ammeters) we might be dealing with direct current sources (d.c.) or alternating current sources (a.c.). It should also be obvious that wide ranges of voltages, currents and resistances may need to be measured, so that the instruments used must have a number of different ranges of measurement.

For many years portable meters with calibrated scales and pointers have been used and are still being extensively employed, but with the advent of cheap electronic circuits we begin to see an increasing use of electronic instruments using a digital display. There is an obvious advantage in the latter type. The voltage, current or resistance is easily read off the display with far less chance of error. There is therefore less skill involved than that required to read off a value from a pointer and scale. There are other advantages too which will be explained later. There is however a distinct disadvantage. An electrical supply is needed to operate a digital meter. There are also relatively complicated pieces of equipment and therefore likely to go wrong, although with the increasing reliability of electronics this last point is becoming of lesser significance. They are also slightly more expensive than the earlier type but the price gap is steadily closing.

Because of the need of measuring three variables, amperes, volts and ohms, one well-known make of indicating test gear incorporating the measurement of all three is called an AVOmeter and has a number of different ranges for current, voltage and resistance, in both d.c. and a.c. circuits. A number of other manufacturers produce somewhat similar meters. Nowadays electronic multirange test meters are available with digital displays as well as the more traditional pointer-type 'AVOs'.

Obviously a number of other electrical quantities may need to be measured; power, frequency, waveform to mention only three. These will not be dealt with at this stage.

Making a measurement

When employing an electrical indicating meter of the type described a number of decisions must be taken, such as:
1. What is the quantity to be measured, ohms, volts or amperes?
2. In the latter two cases is the source direct or alternating?
3. What range is required – milliamperes, amperes, volts or kilovolts?

The meter selected, and the range employed must be decided upon before the meter is applied to the circuit which is being measured. If the wrong range is used damage might ensue to both the indicating meter and the circuit. Other questions should also be going through the mind of the technician when making a measurement.
1. What accuracy is required in making the measurement and does the meter employed provide this accuracy?
2. To what extent does the introduction of the meter disturb the electronic circuit being measured?
3. In the case of alternating quantities, what is the frequency and waveform of the voltage or current and can the meter cope with the particular frequency and waveform?

The last three points are just as important as the first three but unfortunately are often forgotten.

The pointer-type instrument

There are three common types of deflection type instruments: (i) moving-coil; (ii) moving-iron; (iii) electrodynamic.

The first type is probably the most common. It is essentially a d.c. meter but can be adapted to read on both alternating and direct currents. The AVOmeter and many other makes of multirange meters employ a moving-coil movement.

The second type, while being cheap to manufacture, is rarely as accurate or as sensitive as the moving-coil meter.

The third type is usually found in the laboratory situation where a high degree of precision is required. It is also used frequently in the measurement of power in low-frequency circuits.

Each meter requires a deflection torque or twist, i.e. something to cause the point to deflect; a control torque, i.e. something that controls the amount of deflection of the pointer for a given deflecting torque; and a damping torque that prevents the pointer oscillating about its final deflection.

In the moving-coil meter, the deflecting torque is caused by the action of a current flowing in a conductor placed in a magnetic field produced by a permanent magnet. The basic moving part of the moving-coil meter is shown in Fig. 8.1. The current flowing in the coil reacts with the magnetic field generated between the north and south poles of the cylindrically shaped magnet to produce a twisting action (the deflecting torque). This in turn produces a tightening of the control spring (the control torque) and when the deflecting torque exactly balances the control-spring torque the meter gives a steady deflection. The deflection produced at full scale depends upon the current flowing, the strength of the magnetic field and the design of the coil in this field. The bigger the coil and the greater the number of turns, the

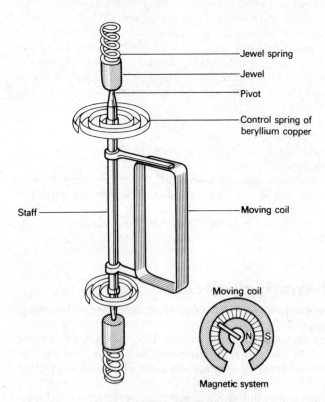

Fig. 8.1 The moving-coil unit

greater the deflection torque for a given current flowing in the coil. The current to cause full-scale deflection also depends upon the strength of the control spring.

Damping is produced by the fact that the coil is wound on an aluminium frame or former.

Moving-coil meters can be made very sensitive but generally speaking the smaller the value of current at full-scale deflection the more delicate they are. This limits their use in a portable situation. A reasonably robust structure with good sensitivity can be obtained with full-scale deflections down to about 50 microamperes (50 μA). The instrument described has a linear scale (one with equally spaced divisions) and the pointer can turn through an angle of up to 300°. Many other designs exist and one producing an angular movement of about 100° is common.

The moving-iron meter has a basic arrangement as shown in Fig. 8.2. The current in the coil magnetises the two pieces of steel so that south poles are adjacent to each other as are the north poles. The result is a repulsion effect (remember like poles repel) and the moving iron attached to the pointer is deflected. The control torque is provided by a spring in precisely the same way as the previously described meter.

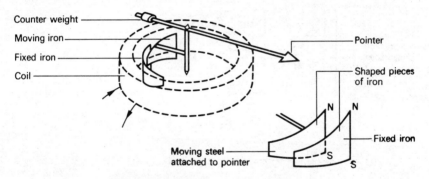

Fig. 8.2 The moving-iron meter

Damping is produced by a 'dashpot' shown in Fig. 8.3. The moving vane is attached to the pointer and as this moves the air in the box is compressed slightly, so cushioning the movement. The amount of damping is affected by the size of the hole in the dashpot. The scale shape is affected by the shape of the pieces of steel which are magnetised. By careful design the scale shape of this type of meter can be made reasonably uniform over about 80 per cent of its length. It gives correct deflection on both direct and alternating currents.

In the electrodynamic instrument (shown basically in Fig. 8.4) the magnetic field produced by the permanent magnet in the case of the moving-coil meter is replaced by an electromagnet (the fixed coils). The deflecting torque is produced by the current in the moving coil reacting with the magnetic field produced by the fixed coils. Damping is again provided by a

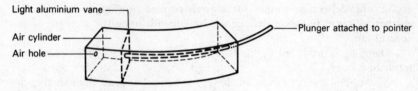

Fig. 8.3 The dashpot

dashpot. The scale shape of this instrument is non-linear, i.e. the spacing of the divisions is unequal. The meter gives correct deflection on both direct and alternating currents.

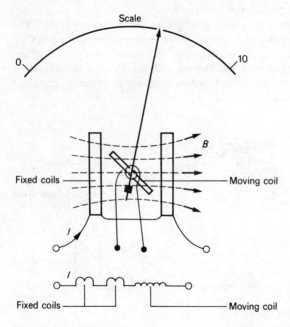

Fig. 8.4 Electrodynamic principle

Ammeters

The deflection produced in the three types of meters described depends directly on the current flowing through them. If the current to be measured exceeds the full-scale deflection for which the meter is designed than a 'shunt' resistor is connected across the meter terminals to divert a proportion of the current being measured from the meter itself (Fig. 8.5). If the shunt diverts $\frac{9}{10}$ of the current leaving $\frac{1}{10}$ to flow through the meter then the effective full-scale deflection is increased ten times, e.g. a meter may have a full-scale

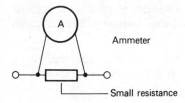

Fig. 8.5 The ammeter

deflection of one milliampere (1 mA). Its resistance (i.e. that of the coil in the moving-coil meter) may be 9 Ω. If now a resistance of 1 Ω is placed across the terminals, $\frac{9}{10}$ of the current flowing to the meter will flow through the shunt. Thus if 10 mA flows to the circuit, 9 mA will flow through the shunt and 1 mA will flow through the meter. 10 mA now causes full-scale deflection, i.e. the reading produced by the meter has to be multiplied by 10. By switching in various values of shunt the range of the meter can be extended by almost any factor. A multirange ammeter is merely one which has a number of different shunts which are switched in turn across the meter.

Voltmeters

There is no essential difference in operation between a voltmeter and an ammeter with pointer-type instruments. A current is needed with either to produce a deflection. In the case of the voltmeter it is necessary to connect a resistor in series with the meter (Fig. 8.6). If the full-scale deflection of the meter is again 1 mA and it possesses a resistance of say 50 Ω then to turn it into a voltmeter having a full-scale deflection of 1 V the total resistance of the circuit must be 1 kΩ (1000 Ω). As the meter already has a resistance of 50 Ω a further series resistance of 950 Ω is necessary. Similarly, to produce a full-scale deflection for 10 V the same meter will require a series resistor of value 9950 Ω. (Why?)

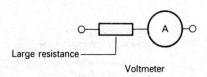

Fig. 8.6 The voltmeter

A range of voltages can be produced using a number of different series resistors. The multirange voltmeter is one which has a number of different resistors switched in turn in series with the meter. The indication has then to be multiplied by a suitable factor.

By suitable switching arrangements it is possible to produce in the one instrument a multirange ammeter and a multirange voltmeter. Typical ranges might be:

100 µA	1 mA	10 mA	100 mA	1 A	(current ranges)
0·1 V	1 V	10 V	100 V	1000 V	(voltage ranges)

Higher voltage ranges may produce insulation and safety problems.

Sensitivity

The voltmeter is inherently an instrument having a large resistance and its sensitivity is expressed in ohms per volt. A voltmeter of 2000 ohms per volt implies that if the full-scale deflection is 1 V then the resistance of the meter is 2000 Ω. This means that the current then flowing is $\frac{1}{2000}$ A, i.e. 0·5 mA or 500 µA. A meter with a sensitivity of 20 kΩ/V has a full-scale deflection when 50 µA flows. The higher the value of ohms/volt the greater the sensitivity of the meter.

Generally speaking the higher the resistance of a voltmeter the better.

The sensitivity of an ammeter is usually expressed as the current required to produce full-scale deflection. Due to the use of shunts, ammeters are inherently low-resistance devices.

Generally speaking the lower the resistance of an ammeter the better.

A.C. ranges

Moving-iron and electrodynamic instruments register correct values on both alternating and direct currents but are rarely used with multirange meters because their sensitivity is significantly less than that of the moving-coil meter. The latter however can be converted to an a.c. meter by using a bridge rectifier (Fig. 8.7) providing full-wave rectification. This allows current to flow through the meter in one direction only, although the voltage applied to

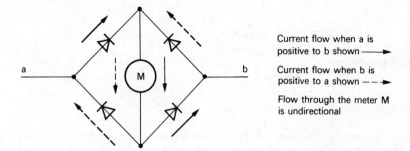

Fig. 8.7 The bridge rectifier

the circuit may be alternating. The meter reads the average value of the current flowing through it but this is not the effective value of the alternating current. It can easily be recalibrated to read the correct value by the adjustment of a suitable shunt or series resistor. With multirange a.c./d.c. meters, the switching from a d.c. range to an a.c. range involves connecting a rectifier and other resistors into the circuit. The calibration of the instrument under a.c. conditions assumes that the waveform of the current or voltage being measured is a sinewave. Although the meter will give an indication for waveforms of a different kind, the deflection is not the effective value of the measured quantity.

In some meters furthermore an indication may still be given if direct voltages or currents are applied when the meter is switched to the appropriate a.c. range. The indication will not usually be the correct value. In the AVOmeter a transformer incorporated in the circuit prevents this happening. If a steady direct current or voltage is applied to the instrument when switched to an a.c. range no indication is given.

The user of the instrument must therefore check carefully not only the range that the instrument is switched to but also whether the meter is connected for a.c. or d.c. measurements. It would be extremely dangerous for example to check whether the 240-V a.c. mains is off with the meter switched to a d.c. voltage range. The fact that no deflection is shown might easily cause the technician to assume (quite wrongly) that the circuit was dead. It could be the technician who became dead in such a situation! The importance of using a multirange a.c./d.c. meter correctly cannot be overstressed.

Ohmmeters

Ohmmeters are based on the principle that a known voltage when applied across a resistor produces a current flow. The application of Ohm's law shows that $R = \dfrac{V}{I}$. Thus the resistance R is proportional to $\dfrac{1}{I}$. The current flowing is therefore proportional to the reciprocal of the resistance — high resistance, small current.

If therefore a battery of known e.m.f. is connected to a suitable ammeter, the deflection of the pointer is dependent upon the resistance of the circuit. The actual arrangement of the ohmmeter is shown in Fig. 8.8. Two alternatives are seen: (i) a series arrangement; (ii) a shunt arrangement. The terminals XX are initially connected together and the zero ohms adjustment is varied until the meter gives full-scale deflection. This is of course the current flowing through the circuit and corresponds to zero ohms across XX. Any resistance now connected in place of the short between XX will cause a reduced current to flow which will correspond to the resistance placed across XX. The adjustment is necessary to allow for changes in the internal resistance of the voltage source. The series arrangement is better for high values of resistance, so that multirange meters usually change the circuit arrangement when switching from a low-resistance range to a high-resistance range.

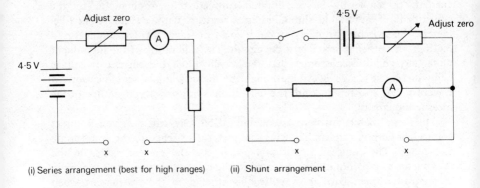

(i) Series arrangement (best for high ranges)   (ii) Shunt arrangement

Fig. 8.8 Ohmmeter circuits

A third method of connection is shown in Fig. 8.9. It should be noticed that the resistance scale has a zero on the right-hand end of the scale, i.e. when the pointer is at full scale. This is the reverse to the situation when measuring either current or voltage when the zero is on the left-hand end of the scale. In multirange meters the battery needed for the resistance ranges is fitted inside the meter. If it is found impossible to 'zero' the ohms range then the internal battery needs replacing.

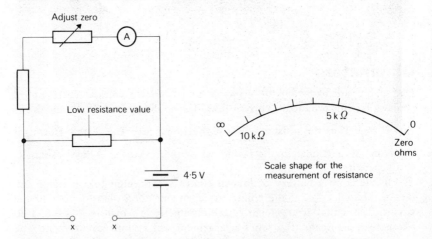

Fig. 8.9 Ohmmeter circuit suitable for measuring low resistance values

The scale of a typical multirange meter is shown in Fig. 8.10a. With the AVO, two range switches are employed — one for direct voltage, current and resistance, one for alternating voltage and current. It should be noted that the resistance scales are non-linear (i.e. very unevenly spaced out).

The complete circuit for the Model 8 AVOmeter is shown in Fig. 8.11.

The electronic multimeter

With the invention of the integrated circuit in large-scale production the cost of electronic equipment has come down drastically and we now have in some instances a situation where the case to put the electronics in, together with

Fig. 8.10a

Fig. 8.10b

the knobs, switches, plugs, etc. are the major items in terms of cost. It is therefore becoming cheaper to design multimeters on an electronic basis rather than use a pointer-type instrument.

The digital meter is such an example and has many advantages in use:
1. It has an easily read display making it far less likely to read the wrong scale.
2. It is probably more sensitive and more accurate than a pointer-type meter.
3. It can withstand electrical overloads and is less likely to damage the circuit with which it is connected when switched to the incorrect range.
4. It requires less skill in use since no estimate of an indicated value is necessary as would happen when the pointer of a meter does not lie precisely over a mark on the scale.

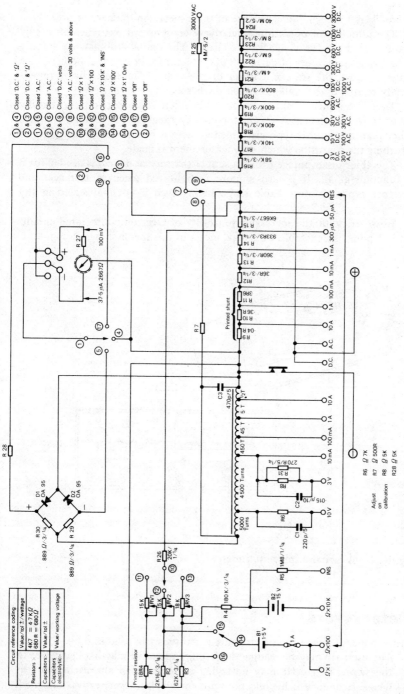

Fig. 8.11 Circuit diagram for Model 8 AVOmeter

We are likely to see an increasing use of these meters in the near future.

The principle of operation is relatively complicated but relies essentially on the time taken for a capacitor to charge. The timing is done electronically by counting the number of cycles of an oscillator (built into the meter) which occurs between the start and finish of the charge. Since the charge time is directly related to the voltage applied the basis of a meter exists.

There is another advantage — especially when measuring voltages. Usually the effective resistance of the meter on any range setting is high — much higher than the pointer-type instrument. There is hence much less chance of disturbing the circuit on which the measurement is made.

The chief disadvantage is that a supply (battery or mains) is needed for it to function at all. It is also a more complicated piece of test gear and therefore probably more liable to go wrong even if not maltreated in any way.

However, with the increasing reliability of electronic integrated circuits this last point is debatable. A typical meter of this sort is shown in Fig. 8.12.

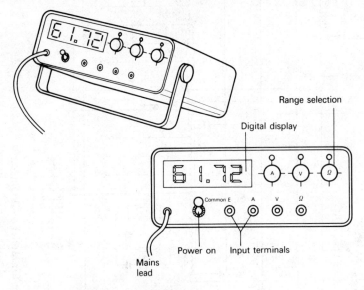

Fig. 8.12   A digital-display multimeter

Meter errors

Errors in measurement can be attributed to three main sources:
1. The user. He may be reading the wrong scale or he may have switched to the wrong range. He may be trying to measure an alternating current whose frequency is beyond the limit for which the meter was designed.

2. **Calibration.** No meter is completely accurate and one has to accept that there are accuracy limits to be taken into account. The d.c. voltage range of the Model 8 AVO is quoted as accurate to within 2 per cent of the indication between full scale and half scale and below half scale accurate to within 1 per cent of full-scale deflection. Unless a shunt or a series resistor has changed its value (possibly due to overloading) or some part of the circuit is not functioning correctly, the calibration error is usually negligible. It is seldom necessary to be very precise in many practical instances. Where high precision is needed then usually the multirange meter is replaced by more sophisticated (and expensive) equipment.
3. **Systematic error.** This is due to the fact that the introduction of a meter into a circuit disturbs the circuit being measured. While it may be possible to make allowance for the disturbance, often the technician may fail to realise that the meter is giving the wrong reading. The most frequent occurrence of this source of error is the measurement of voltage in a circuit containing high values of resistance. Under these conditions the meter resistance may be comparable with that of the circuit.

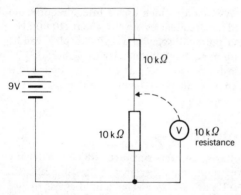

Fig. 8.13 (Example)

As an example consider the circuit in Fig. 8.13. Since there are 9 V across the two equal resistors the voltage across one of them is 4·5 V. If however a voltmeter having a resistance of 10 000 Ω is placed across one resistance, the effective resistance of the parallel combination is now only 5000 Ω. The p.d. across 5000 Ω is therefore ⅓ of the total voltage since a 10 000-Ω resistor is still in series. The meter registers not 4·5 V but one-third of 9 V, i.e. 3 V. As a percentage of the indication the meter is reading low by 1·5 V in 3 V, i.e. 50 per cent error. Switching to a less sensitive range will increase the voltmeter resistance and hence reduce this source of error. Unfortunately the deflection of the pointer is also reduced so that it is liable to produce an observational error which again can be a fairly high percentage of the indicated value.

An electronic meter on the other hand may have a resistance well in excess of 1 MΩ. If such a meter is used in this case the disturbance effect is much less (actually with the figures given around 1 per cent). In making voltage measurements in high resistance circuits an electronic meter (whether or not a digital display is employed) is usually advantageous.

Frequency response of indicating instruments

Indicating instruments (ammeters and voltmeters) have a frequency operating range outside which the calibration can no longer be relied on. The moving-iron and electrodynamic meters are essentially low-frequency devices and cannot be employed to measure currents or voltages above a few hundred hertz without considerable error.

The moving-coil meter, which is basically a d.c. instrument, can be made to indicate on alternating current by using a bridge rectifier. It is then the frequency limit of the rectifier which imposes the maximum frequency at which the meter can be used. It should also be pointed out that the calibration holds only if the waveform of the current or voltage being measured is a sinewave.

The multimeter such as the AVOmeter which uses a bridge rectifier on the a.c. ranges can be employed at frequencies up to about 20 000 Hz. Electronic meters, whether using a pointer display or a digital display, can be used at much higher frequencies, up to several million hertz (megahertz).

The selection of the right instrument for a particular measurement is therefore influenced by the frequency of the signal being considered.

Bandwidth measurements

Amplifiers are used to enlarge voltages, currents or power which can occupy very wide frequency bands. An audio amplifier may have to deal with frequencies ranging from about 50 Hz to 25 kHz; a video amplifier has to deal with T.V. signals whose frequency may extend to 5 MHz; a radio receiver may have to deal with frequencies up to 50 times this figure. The electronic type of indicating meter is the only type which can be used to measure amplifier characteristics over such a wide frequency range.

The testing of an amplifier (whether audio, video or radio) requires a variable-frequency source, called a signal generator, capable of producing voltage sinewaves over the frequency range of the amplifier. The magnitude of the output of the generator must also be variable and measurable. The usual form of signal generator is a variable-frequency oscillator with a number of alternative ranges of frequency. The output is applied to an attenuator circuit which allows a known portion of the measured output voltage to be applied to the circuit under test.

A typical block diagram showing the test circuit is shown in Fig. 8.14. The frequency of the signal generator is set at the lower limit at which the amplifier functions. The attenuator is adjusted to give a known magnitude of output voltage, which is then applied to the input of the amplifier, care being taken not to exceed the maximum value for which the amplifier has been designed. The output of the amplifier across the load resistor R_L is measured on a voltmeter which has a frequency operating range at least as wide as the

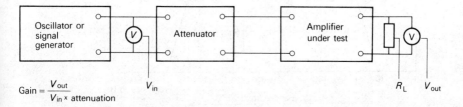

$$\text{Gain} = \frac{V_{out}}{V_{in} \times \text{attenuation}}$$

Fig. 8.14 Bandwidth test circuit

amplifier under test. The output voltage is then measured at a number of frequencies keeping the magnitude of the input voltage constant. The gain of the amplifier is the ratio of the output voltage to the input voltage, and the shape of the graph depends upon the type of amplifier being tested.

Two frequency responses (an audio amplifier and an tuned radio frequency amplifier) are shown in Fig. 8.15. The horizontal axis of the graph uses equal intervals corresponding to frequencies having powers of 10, e.g. 10, 100, 1000, etc. This is called a logarithmic scale of frequency. The bandwidth of an amplifier is the frequency range over which the gain is equal to or greater than 0·71 × maximum gain.

An audio amplifier is designed to have a uniform gain over a wide frequency range. Tuned radio-frequency amplifiers are designed to amplify voltages or currents over a much smaller frequency range (at least expressed as a percentage of the centre frequency).

Manufacturer's specifications

Amplifiers may be sold as having a certain gain over a given bandwidth. The term **decibel** (abbreviated to dB) is normally used in such specifications. When the term is employed in defining the frequency response of amplifiers the decibel can be expressed as follows:—
The fall in gain at some frequency f is

$$20 \log_{10} \left(\frac{\text{maximum gain}}{\text{gain at frequency } f} \right)$$

Thus if the gain is 1000 times at 5 kHz, this being the maximum gain, and the gain at 25 Hz falls to 100, then the fall in gain expressed in decibels is

$$20 \log_{10} \frac{1000}{100} = 20 \log_{10} 10$$

But $\log_{10} 10 = 1$.
Hence the fall in gain is 20 dB.

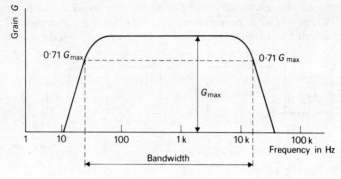

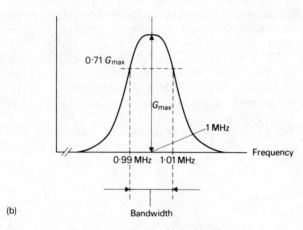

Fig. 8.15 Radio frequency tuned amplifier response

Similarly, if the gain at some other frequency falls to 710 the fall in gain is

$$20 \log_{10} \frac{1000}{710} = 20 \log_{10} 1.428 = 20 \times 0.15 \text{ approx.}$$
$$= 3 \text{ dB.}$$

Since the bandwidth of an amplifier can be defined as the frequency range over which the gain does not fall by more than approximately 0·71 of the maximum gain, the bandwidth can also be defined as the frequency range over which the gain does not fall by more than 3 dB from its maximum value. The decibel is used extensively in defining the response of electronic circuits.

The more precise definition of the decibel is as follows, since the unit is essentially a power ratio.

Power ratio in decibels $= 10 \log_{10} \dfrac{\text{output power}}{\text{input power}}$

$= 10 \log_{10} \dfrac{(\text{output voltage})^2/\text{output resistance}}{(\text{input voltage})^2/\text{input resistance}}$

Where the input resistance and the output resistance have the same value the power ratio in decibels is given by $10 \log_{10} \dfrac{(\text{output voltage})^2}{(\text{input voltage})^2}$

$= 10 \log_{10} \left(\dfrac{\text{output voltage}}{\text{input voltage}}\right)^2$

$= 20 \log_{10} \dfrac{\text{output voltage}}{\text{input voltage}}$

since $\log_{10} A^2 = 2 \log_{10} A$.

On this basis Table 8.1 gives some ratios for the gain of an amplifier.

Table 8.1.

Decibels	Gain falls to this value of its maximum
0.1	99%
0.5	94%
1.0	89%
2.0	79%
3.0	71%
4.0	63%
5.0	58%
6.0	50%
10	32%
15	18%
20	10%
50	0.3%
100	0.001%

Questions

Indicate by a T (true) or F (false) which of the following statements are correct. In the case of false statements, give the correct statement.
1. Moving-iron meters give the correct value when measuring either low-frequency alternating or direct current.
2. Moving-coil meters can only be used in d.c. measurements.
3. Damping in a moving-coil meter is provided by a dashpot.

4. Ammeters are essentially high-resistance devices.
5. Voltmeters employ resistance in series with the meter movement.
6. Sensitivity with voltmeters is usually expressed in ohms/volt.
7. Ohmmeters usually have non-linear scales with the zero on the right-hand end of the scale.
8. The electrodynamic instrument is frequently employed in multimeters.
9. A bridge rectifier provides full-wave rectification.
10. The moving-coil meter usually provides a non-linear scale when used as a d.c. voltmeter.
11. The resistance of all multimeters remains constant when switching from one range to another.
12. All electronic multimeters have digital displays.
13. The moving-iron meter cannot be used at radio frequencies.
14. Electronic indicating meters cannot be employed in circuits operating at mains frequencies.
15. Amplifiers always have a uniform gain over a wide frequency range.
16. The decibel scale expresses a power ratio.

Exercises

1. A meter having a full-scale deflection (F.S.D.) of 250 V is liable to a calibration error of ± 2 per cent of F.S.D. The observational error is ± 1 per cent of F.S.D.
   If the meter indicates 120 V between what limits does the voltage lie?
   Ans. 127·5 V, 112·5 V.

2. Find the indication on the voltmeter in the circuit shown in Prob. Fig. 8.1 when switched to:
   (i) 1-V range; (ii) 10-V range; (iii) 100-V range.

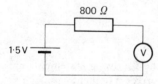

Prob. Fig. 8.1

The instrument has a sensitivity of 1000 Ω/V. (Calibration and observational errors can be neglected.)
Ans. 0·83 V, 1·39 V (1·4 V), 1·49 V (1·5 V).

3. A Model 8 Universal AVOmeter switched to measure current gives a reading on both a.c. and d.c. scales (although not the same). Account for this.

4. A meter has a resistance of 10 Ω and a full-scale deflection of 1 mA. How can it be adapted to read voltages up to 100 V and currents up to 1 A? Find in each case the value of any additional components needed.

5. Define the term 'bandwidth' when applied to (a) an audio amplifier used in a 'hi-fi' system; (b) a tuned radio frequency amplifier.
   Illustrate your answer by suitable graphs.

6. Define the term decibel.
   An amplifier gain falls by 3 dB from a figure of 700. What is the new gain?
   Ans. Approx. 500.

7. Find the indication on the Voltmeter V connected to the circuit shown in Prob. Fig. 8.2 if the sensitivity of the meter is 200 Ω/V and it is switched: (a) to the 100-V range; (b) to the 50-V range.

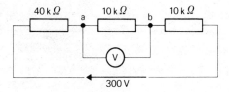

Prob. Fig. 8.2

Calibration error may be ignored. What is the true value of V_{ab}?
Ans. 35·3 V, 18·6 V, 50 V.

8. What is the current flowing in each branch of the circuit in Prob. Fig. 8.3?

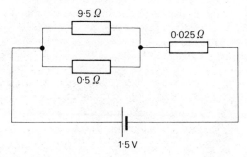

Prob. Fig. 8.3

What will an ammeter of resistance 0·5 Ω and full-scale deflection 3 A register if inserted into each branch?
Ans. 3 A, 2·85 A, 0·15 A, 1·5 A, 1·46 A, 0·14 A.

7. A resistance R is measured using a voltmeter and ammeter as shown in Prob. Fig. 8.4. If V registers 52 V and A registers 480 mA, what is the apparent value of R?

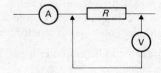

Prob. Fig. 8.4

If the resistance of the voltmeter is only 1000 Ω, what is the true value of R, ignoring any calibration error?
Ans. 108·3 Ω, 121·5 Ω.

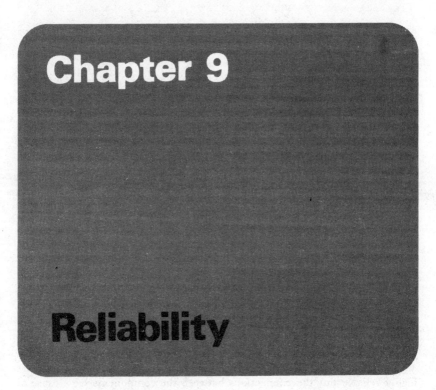

Chapter 9
Reliability

Concept of reliability

Reliability of electronic systems is concerned with their satisfactory performance, that is, systems which do not break down. At first glace it would seem that we need systems which do not break down at all but this implies complete reliability, perfect components and infinite lifetimes. It is arguable if it makes economic sense to aim at such perfection. Systems become obsolete after a time and are replaced by newer and possibly better systems, so the need to make electronic systems continue indefinitely may not necessarily be a sensible objective. Furthermore, making the components of a system as near perfect as possible costs money and may consequently push up the price of a system beyond that which the market is prepared to pay.

Obviously there are systems where long life or high reliability are very desirable features — the life-support system on a spacecraft or the heart pacemaker are two obvious examples.

Another case where a high order of reliability is needed is the under-ocean telephone repeater which would firstly cost a lot of money to raise from the ocean bed to repair, and secondly would interrupt an international communication system while it was out of action.

Hence high reliability must be assessed against the situation in which the equipment is being employed and the cost of making it reliable.

At the other end of the scale no one wants unreliable systems even at a low cost.

Failure of a system

Even the term 'failure' needs to be somewhat qualified. Failure of an entire system is rare — normally only component or sectional failures occur which may or may not render the whole system useless.

If we take as an example a car, we can have a failure in the ignition system and the car cannot function, although the fault may easily be put right if it is merely a broken electrical lead. On the other hand a broken suspension spring or even a flat tyre does not render a car useless. We can still travel (admittedly more carefully) with the former and we now have specially designed tyres on which even though badly punctured, we can travel at fairly high speeds for perhaps a hundred or more kilometres. The cost of repair in these cases however is significantly greater than in the case of the broken lead. Electronic systems can be very complicated and we may find partial or complete breakdowns which put the system out of service or we may be able to continue to use a system at reduced efficiency even though part of it is non-functional. Consequently when we talk about breakdown it is necessary to define what we mean.

Often in semiconductor circuits the components within the circuit are highly reliable but the connections between the components are a cause of breakdown. Breakdown too may be due to maltreatment of equipment. If too high a voltage is applied to a circuit, for example, or the polarity of a battery is incorrect, we must not blame the equipment if it ceases to function correctly.

Usually there are limits imposed on the reliability by such errors of the user and should any electronic system break down or fail to operate we must first check that we are not in any way contravening specific instructions issued with such apparatus.

Nevertheless one has to face up to the fact that failures do occur even with careful design and close supervision in manufacture and it is necessary to use such phrases as mean time to failure (M.T.T.F.) or mean time between failures (M.T.B.F.). The second expression implies that the system is readily repairable. The M.T.T.F. also affects the guarantee which a manufacturer is prepared to place on equipment.

Lifetime measurements

Let us take the case of electric filament lamps. Now although it is possible to produce 'long-life' lamps there is unfortunately a tendency for such lamps to be slightly less efficient than other lamps. There is also a need to find out whether it is cheaper to replace the more efficient bulbs more frequently since long-life bulbs inevitably cost more to produce. It is necessary also to test the lifetimes of lamps in order that the manufacturer can monitor the quality of the lamps and guarantee his product. This means switching lamps on and finding out how long they last before burning out.

Five filament lamps tested in this way may have, for example, lives of 1200 hours, 1500 hours, 1150 hours, 1900 hours and 1350 hours. The average life of these five lamps is

$$\frac{1200 + 1500 + 1150 + 1900 + 1350}{5} = 1420 \text{ hours}$$

If the sample chosen is typical of all the lamps manufactured we could say that the average life of the lamps was about 1400 hours, i.e. the M.T.T.F. for the sample of five lamps was 1400 hours. The manufacturer of the lamps would be foolish to guarantee such a life time since on average 50 per cent of the lamps would fail before 1400 hours and would have to be replaced free of charge.

He could, however, guarantee the lamps to, say, 1000 hours as none of the samples had a shorter life. Of course there is no assurance that the lamps he has tested are typical of the whole of the production but at least one would expect that the number failing before the end of such a guaranteed period would be small. One has to take a risk in this sort of situation since lifetime tests destroy the product. There are certain mathematical rules derived from statistical theory on which the number selected for testing is based and on which the risk factor can be estimated, but this is beyond the scope of this book.

The last example also raises another point. It may take a considerable time before certain types of equipment fail and the manufacturer is often not prepared to wait too long for the completion of a life time test. With filament lamps the testing situation may be made such that the lifetime of the lamps is artificially decreased. This can be done by deliberately applying a higher voltage than that for which they are designed and switching the lamps on and off at perhaps one-second intervals. Switching causes current surges which do not occur with apparatus in continuous operation, hence such action shortens the life of electrical and electronic equipment.

Failure rate

Failure rate is an expression frequently employed when reliability is concerned. It can be expressed as the number of component failures per hour or per thousand hours. It is implied that the failure rate applies to equipment being employed under its normal rated conditions. The failure rate f is related to the M.T.T.F. (T) by the simple relationship

$$f = \frac{1}{T}$$

Thus if the M.T.T.F. is 250 hours the failure rate is

$$\frac{1}{250} = 0.004 \text{ per hour or 4 per 1000 hours.}$$

Equipment which has been only recently developed usually has a higher failure rate than equipment which has been available for some time. This is because there are usually hidden weaknesses in a new model which become apparent only after the equipment has been in use (or abuse) by 'inexperts' or by persons unfamiliar with it. The manufacturer is continually making improvements or alterations and tries to ensure that as far as possible all the defects are removed before selling a product to the general public. He may in fact choose to put a limited number of a new product in the hands of a few persons who might reveal initial weaknesses and report these to him at the end of a time period. The failure rate consequently falls during this initial 'burn in' period. After this initial phase the failure rate should remain constant, breakdowns being of a purely random and unpredictable nature.

The final phase is where a product tends to wear out due to continual use. A thermionic diode, for example, may have a cathode surface which begins to lose its electron emission quality after prolonged use. The failure rate rises once more during the 'wear out' or 'burn out' phase. The total graph of failure rate against time shows a shape similar to that shown in Fig. 9.1. It is called a 'bath-tub' curve due to its shape. The manufacturer has to decide: (i) whether the flat portion of the graph occupies sufficient time, i.e. whether the product is reliable over a long enough period; and (ii) whether the failure rate during this period is low enough. Two possibly unacceptable 'bath-tub' curves are illustrated in Fig. 9.2.

Linked systems

Systems which are in series with one another have a higher failure rate than systems in parallel (Fig. 9.3).

Series systems each with a M.T.T.F. of T_1, T_2, T_3, etc., have a failure rate f given by

$$f = \frac{1}{T_1} + \frac{1}{T_2} + \frac{1}{T_3} \text{ etc.}$$

Thus if $T_1 = 1000$ hours, $T_2 = 500$ hours, $T_3 = 250$ hours, the failure rates of each separate system are:

$$f_1 = \frac{1}{T_1} = 0.001 \text{ per hour}$$

$$f_2 = \frac{1}{T_2} = 0.002 \text{ per hour}$$

$$f_3 = \frac{1}{T_3} = 0.004 \text{ per hour}$$

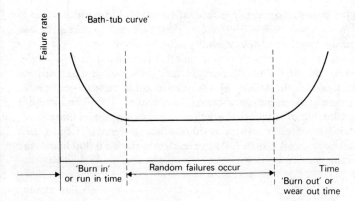

Fig. 9.1 'Bath-tub curve'

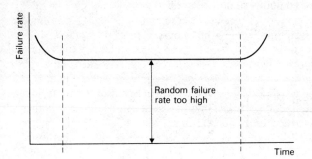

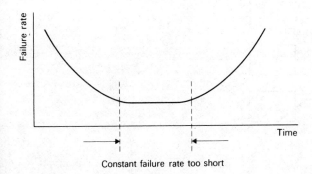

Fig. 9.2 Two unacceptable graphs

f for the whole series arrangement = 0·001 + 0·002 + 0·004 = 0·007 per hour. since any one system causes breakdown.

The M.T.T.F. = $\dfrac{1}{0\cdot 007}$ = 143 hours

This is much lower than that of any individual system and shows the effect of one section with a low M.T.T.F. Thus the weakest link in the chain is the one responsible for the lowering of the reliability of the system.

Components in an electronic circuit may be considered to be in series when calculating the M.T.T.F. although in fact they may not actually be connected in series, if the failure of any one of them causes total circuit failure. For example, a circuit may contain 6 resistors, 4 capacitors and 2 transistors, breakdown of any one of which may produce system failure.

If the M.T.T.F. of each resistor is 20 000 hours, the M.T.T.F. of each capacitor 4000 hours, and the M.T.T.F. of each transistor 10 000 hours, the failure rate of the system is

$$f = \frac{6}{20\,000} + \frac{4}{4000} + \frac{2}{10\,000}$$

$$= \frac{3}{10\,000} + \frac{10}{10\,000} + \frac{2}{10\,000} = \frac{15}{10\,000} \text{ per hour.}$$

The M.T.T.F. of the system is $\frac{10\,000}{15} = 667$ hours.

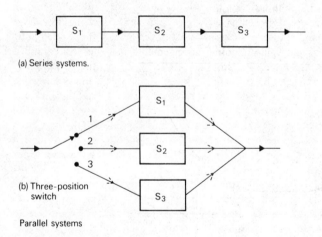

Fig. 9.3 Parallel systems

Highly reliable systems are sometimes built with two or more parallel systems so that in the event of failure of one system the parallel system can be switched in. Such arrangements are obviously more expensive since part of the system is out of use all the while the main system functions. Aircraft systems, where a failure might prove catastrophic, have often two or even three parallel arrangements for increased reliability.

Factors affecting failure

Switching equipment on and off frequently usually shortens the M.T.T.F. and this should be borne in mind when using electronic apparatus. Complex equipment — for example a computer — is better left permanently on rather than being switched on and off frequently if a high degree of reliability is required. Furthermore, a regular maintenance schedule (possibly occupying 1 hour per day) may reveal subunits or components which are nearing their limit of useful life before they actually break down, and enables these to be replaced. It is now becoming possible to feed programmes into computers or other systems which detect automatically, and very quickly, suspect units or components. Preventive maintenance, as this is called, means additional expense for two reasons: (i) the cost of the technician's time or the cost of the automatic test equipment; and (ii) the fact that during this period the equipment is not available to the normal user. Nevertheless such an arrangement may be very acceptable to the user. He knows, for example, that by losing 1 hour of computing time per day he has a much greater chance of uninterrupted use of trouble-free equipment for the rest of the working day. He loses no working time at all if the preventive maintenance is carried out during the time the equipment is not normally used, i.e. at night.

Another factor which affects failure rate is the overstressing of components. In much the same way as a bridge is designed to carry a safe load, an electronic component is designed to carry a safe current or have a safe voltage applied. Resistors and transistors especially can break down if their temperature exceeds a designed value. Cooling arrangements may be needed and if these fail (e.g. a motor-driven fan stops) then the system is much more likely to give trouble. Capacitors have a safe working voltage and if this is exceeded the insulation is liable to break down. Sometimes the breakdown of one component causes an adjacent component to fail and the effect can be disastrous. Capacitors in series can produce a cumulative breakdown if each has its full rated voltage across the plates. Shorting together of one pair of plates by breakdown then transfers an even greater stress to the other capacitors in the link.

Thermionic-valve heaters on the other hand must have the full voltage applied in order that the cathode can reach its proper operating temperature. Prolonged use of a thermionic device with too low a cathode temperature may in fact cause rather more damage than a small overload of heater current (say 10 per cent). The manufacturer's specifications must be adhered to if reliable systems are to be produced.

Repairable systems

Failures with computers should of course be rare but when such a system has failed it is usually repairable, often quite quickly, and the term to use is mean time between failure (M.T.B.F.).

The failure rate in such cases is related to the M.T.B.F. by the same relationship as before

$$\text{failure rate} = \frac{1}{\text{M.T.B.F.}}$$

It should be appreciated that in an electronic system it is both unlikely to happen and unnecessary to ensure that each component in it has the same failure rate. It is known that certain components are more likely to fail than others. Valves are, generally speaking, more liable to cause trouble after, say, 1000 hours working than transistors with the same operating span. Mechanical components, e.g. contacts and relays, are more liable to break down that electrical components, e.g. resistors, chokes, etc. It is therefore necessary to concentrate on the areas of weakness to design reliable systems. If it is decided that taking all things into consideration a certain failure rate is to be tolerated then a good designer concentrates on making those sections of the system easily accessible and easily replaceable. It is often better to arrange electronic systems as a series of plug-in units so that the faulty component can be quickly located and a replacement easily fitted. Increasingly, the technician's job is to locate faults and replace subsystems rather than locate faulty components and replace them.

Summary

Need for reliability in a system
Mean time to failure (M.T.T.F.). Mean time between failures (M.T.B.F.). Failure rate. Lifetime measurements. Series and parallel arrangements. Factors affecting lifetime. Preventive maintenance. Design considerations.

Questions

1. An electronic circuit has 5 capacitors, 20 resistors and 3 transistors. The M.T.T.F. of the capacitors is 5000 hours and of the resistors 10 000 hours. The failure rate of the transistors is 0·0004 per hour. If failure of any one component means the system is non-functional, what is the M.T.T.F. of the circuit?
What is the most reliable component?
Ans. 238 hours.

2. What component is most liable to breakdown in the previous question and what is the M.T.T.F. required of this component if the M.T.T.F. of the system is to be 250 hours?
Ans. 3750 hours.

3. 10 000 components have a constant failure rate and every day 10 per cent of the serviceable components fail. Draw a graph showing how the number of serviceable components vary over a period of 10 days. What time elapses before only 5000 components are serviceable?

Index

active component, 121
alkaline cell, 89
alphanumeric code, 71
ammeter, 136
ampere, 32, 133
   hour, 92
amplifier, 49
   linear, 50
analogue signal, 60
AND statement, 126
attenuator, 49
AVOmeter, 133

Baird, 6
bandwidth, 65, 146
bath-tub curve, 156
B.B.C., 6, 25
Bell, 10
bias, 4
   forward, 81
   reverse, 81
binary
   code, 71
   signal, 61
bipolar transistor, 93
bit, 61
bit rate, 61
black box, 47, 48
block diagram, 50
bridge rectifier, 87

capacitance, 109
capacitor, 100, 109
   electrolytic, 112
   mica, 112
   paper, 111
   smoothing, 84
   reservoir, 84

carrier, 66
   majority charge, 80
   minority charge, 80
cell
   alkaline, 91
   lead acid, 90
   Leclanché, 98
   mercury, 89
   primary, 88
   secondary, 89
channel, 68
choke, 84
code
   alphanumeric, 71
   binary, 61
   colour, 106
   error detecting, 74
coding, 60
conductor, 40
connection
   parallel, 36
   series, 35
connectors, 121
coulomb, 92
current, electric, 2, 32

decibel, 147
deForest, 4
demodulation, 69
denary, 70
dielectric, 103
digital
   meter, 142
   signal, 61, 69
diode
   semiconductor, 80
   thermionic, 3, 78
distortion, 49

Edison, 2
   effect, 4
electric current, 2, 32
   shock, 28
electrodynamic, 136
electron, 1, 3
electronics, 1
   industry, 20
   system, 47
ENIAC, 8
error detecting code, 74
errors
   meter, 144
   systematic, 145
   transmission, 73
EXCLUSIVE OR statement, 126

facsimile, 65
failure, 154
   rate, 155
farad, 110
field effect transistor, 6
Fleming, 3
forward bias, 81
frequency, 62
   division multiplexing, 66
full-wave rectification, 83

G.E.C., 20
Gate-NAND, 50
half-wave rectification, 83
heat sink, 120
henny, 110
Hertz, 10, 62

I.E.E., 12
I.E.R.E., 12
inductor, 100, 106
industry, electronics, 20
information, 59
Institute, Royal, 20
input impedance, 96
internal resistance, 92

laser, 8
lead and cell, 89
Leclanché cell, 88
linear amplifier, 50
   resistance, 40
logic symbols, 127, 128

Marconi, 11
   Company, 18

Maxwell, 9
mercury cell, 89
meters
   error, 144
   moving coil, 134
   moving iron, 135
microcircuit, 52
modulation, 66
Morse, 9
MTBF, 154
MTTF, 154
multiples, 102

n-type semiconductor, 80
NAND gate, 50
non-linear resistance, 121
NOT statement, 127
npn amplifier, 95

ohm, 33, 133
ohmmeter, 139
Ohm's law, 32
OR statement, 125
output impedance, 96

p-type semiconductor, 80
parallel connection, 36
parity bit, 74
peak value, 62, 64
planar transistor, 93
Post Office, 17, 20
potential difference, 32
power
   gain, 97
   rating, 104
primary cell, 88
printed circuit, 54, 56
pulses, 64

radar, 7
radian, 63
rectifier-selenium, 5
rectification
   full wave, 83
   half wave, 82
reliability, 153
reservoir capacitor, 84
resistance, 33
   linear, 40
   non-linear, 41
resistors, 100
reverse bias, 81
ripple voltage, 84

Royal Institute, 11
Royal Society, 11

safety, 26
secondary cell, 89
semiconductor, 40, 80
   diode, 80
   n-type, 80
   p-type, 80
sensitivity, 138
series connection, 35
shock, electric, 28
shunt, 136
side band, 67
signal
   analogue, 60
   digital, 61, 69
   T.V., 65
sinewave, 62
smoothing capacitor, 84
Society, Royal, 65
symbols
   FET, 97
   logic, 127, 128
   resistor, 34, 106
   transistor, 94, 122
system, 46
   electronic, 47

systematic error, 145

temperature co-efficient, 40
thermionic
   diode, 78
   triode, 4
   valve, 3
Thomson, 2
transfer function, 49
transformer, 82
transistor, 5
   biopolar, 93
   field effect, 6, 97
   planar, 93
   symbol, 94
transmission errors, 73
triode, 4
T.V., 6
signal, 65

valve, thermionic, 3
volt, 32, 133
volt meter, 137

Watt, 103
Wheatstone, 9